International Political Economy Series

Series editor
Timothy M. Shaw
Visiting Professor
University of Massachusetts Boston, USA

and

Emeritus Professor
University of London, UK

The global political economy is in flux as a series of cumulative crises impacts its organization and governance. The IPE series has tracked its development in both analysis and structure over the last three decades. It has always had a concentration on the global South. Now the South increasingly challenges the North as the centre of development, also reflected in a growing number of submissions and publications on indebted Eurozone economies in Southern Europe. An indispensable resource for scholars and researchers, the series examines a variety of capitalisms and connections by focusing on emerging economies, companies and sectors, debates and policies. It informs diverse policy communities as the established trans-Atlantic North declines and 'the rest', especially the BRICS, rise.

More information about this series at
http://www.springer.com/series/13996

Karen M. Siegel

Regional Environmental Cooperation in South America

Processes, Drivers and Constraints

Karen M. Siegel
School of Social and Political Sciences
University of Glasgow
Glasgow, UK

International Political Economy Series
ISBN 978-1-137-55873-2   ISBN 978-1-137-55874-9   (eBook)
DOI 10.1057/978-1-137-55874-9

Library of Congress Control Number: 2017939325

This Palgrave Macmillan imprint is published by Springer Nature
The registered company is Macmillan Publishers Ltd.
The registered company address is: The Campus, 4 Crinan Street, London, N1 9XW, United Kingdom

ACKNOWLEDGEMENTS

I have been very lucky to have the support of many people during the work on this project. Mo Hume and Kelly Kollman have provided invaluable feedback and support throughout the research and writing process. I would also like to thank Jean Grugel for her insightful comments and encouragement. I am also grateful to have had the observations and suggestions of the late Nicole Bourque; her enthusiasm, laughter and willingness to listen and help others are greatly missed. Thanks also to all the people in Politics at Glasgow University for providing a friendly and supportive research environment. Special thanks to Koen Bartels who has been an enthusiastic discussion partner on all things related to methodology. In addition, I would like to acknowledge the Earth System Governance Project and in particular the participants at the Lund Conference in 2012 and the Nairobi Conference in 2016 who provided helpful feedback on parts of the findings. Moreover, I would like to thank Timothy Shaw and the two anonymous reviewers for their helpful comments and suggestions and Christina Brian for her help during the publication process. Thanks also to the University of Glasgow, formerly the School of Law, Business and Social Sciences for generous funding for this project and the Society for Latin American Studies which has provided grants for fieldwork and conferences.

Thanks are also due to numerous people on the other side of the Atlantic. It would not have been possible to write this book without the help and time offered by all the people interviewed for this project and the many people who have helped in making contacts for interviews, sourcing information or engaged in informal discussions. Of course, the interpretation of the interview data and any errors are my responsibility. I would also like to thank FLACSO Argentina for hosting me as a guest researcher during my fieldwork. Special thanks to Diana Tussie, Mercedes Botto, Gastón Fulquet and Marcelo Saguier for many interesting discussions and making me feel welcome in Buenos Aires.

Finally, I have to thank my friends and family for all their support over the last years. I have always been able to rely on my parents for help and encouragement. Andrina and David have been extremely generous in lending us their car to escape from the city once in a while which allowed me to come back to the writing with a fresh mind. At last, I'd like to thank my partner Greg who has made a very tangible contribution by dealing with all the IT issues arising during this project, from encrypting interview data to finding the best solutions for editing and referencing. But more importantly, he has always been there for me, whether I was in Scotland or South America. I dedicate this book to him.

CONTENTS

About the Author

Karen M. Siegel is a Lord Kelvin Adam Smith Research Fellow at the University of Glasgow, UK. She has been a Visiting Researcher at the Facultad Latinoamericana de Ciencias Sociales (FLACSO) Argentina, and she is an elected member of the Society for Latin American Studies (SLAS) Committee as well as a Research Fellow of the Earth System Governance Project.

ABBREVIATIONS

Alba   *Alianza Bolivariana para los Pueblos de nuestra América*
                (Bolivarian Alliance for the Peoples of our America)
ASEAN   Association of Southeast Asian Nations
CEFIR   *Centro de Formación para la Integración Regional* (Centre of
                Education for Regional Integration)
CIC   *Comité Intergubernamental Coordinador de los Países de la Cuenca
                del Plata* (Intergovernmental Coordinating Committee of the La
                Plata Basin Countries)
CLAES   *Centro Latinoamericano de Ecología Social* (Latin American Centre
                of Social Ecology)
CMS   Convention on Migratory Species
COMIP   *Comisión Mixta Paraguayo-Argentina del Río Paraná*
                (Paraguayan-Argentinean Joint Commission of the Paraná River)
COSIPLAN   *Consejo Suramericano de Infraestructura y Planeamiento* (South
                American Council of Infrastructure and Planning)
CyMA   *Competitividad y Medio Ambiente* (Mercosur project
                Competitiveness and Environment)
ECLAC   Economic Commission of Latin America and the Caribbean
EU   European Union
FAO   United Nations Food and Agriculture Organisation
FARN   *Fundación Ambiente y Recursos Naturales* (Environment and
                Natural Resources Foundation)
FLACSO   *Facultad Latinoamericana de Ciencias Sociales* (Latin American
                Social Sciences Institute)
FONPLATA   *Fondo Financiero para el Desarrollo de la Cuenca del Plata*
                (Development Fund of the La Plata Basin)
FTAA   Free Trade Area of the Americas

GCFA	*Grupo de Conservación Flamencos Altoandinos* (High Andes Flamingo Conservation Group)
GEF	Global Environment Facility
GM	Genetically-Modified
GIZ	*Deutsche Gesellschaft für Internationale Zusammenarbeit* (German agency for international cooperation)
GTZ	*Deutsche Gesellschaft für Technische Zusammenarbeit* (German agency for technical cooperation)
ICJ	International Court of Justice
IDB	Inter-American Development Bank
IIRSA	*Iniciativa para la Integración de la Infraestructura Regional Suramericana* (Initiative for the Integration of Regional Infrastructure in South America)
IMF	International Monetary Fund
IUCN	International Union for Conservation of Nature
Mercosur	*Mercado Común del Súr* (Common Market of the South)
MST	*Movimento Sem Terra* (Landless Workers Movement)
NAFTA	North American Free Trade Agreement
NGO	Non-Governmental Organisation
OAS	Organization of American States
REMA	*Reunión Especializada de Medio Ambiente* (Specialised Meeting on the Environment)
SGT	*Subgrupo de Trabajo* (Mercosur working subgroup)
SGT6	*Subgrupo de Trabajo 6 Medio Ambiente* (Mercosur working subgroup 6 on the environment)
Unasur	*Unión de Naciones Suramericanas* (Union of South American Nations)
UN	United Nations
UNCCD	United Nations Convention to Combat Desertification
UNDP	United Nations Development Programme
UNEP	United Nations Environment Programme
US	United States (of America)

LIST OF FIGURES AND TABLES

CHAPTER 1

Introduction

In June 1992, two-thirds of the world's heads of state and thousands of non-governmental organisation (NGO) representatives met in Rio de Janeiro in Brazil for the United Nations (UN) Conference on Environment and Development or Earth Summit. The event was remarkable for several reasons. Never before had so many countries sent their highest representatives to discuss how to address shared environmental problems, an issue which generally does not make it to the top of government agendas. The summit was also noteworthy for its location. It clearly demonstrated that important changes had recently taken place in Brazil and in the Southern Cone of South America more generally. During the 1980s, all the countries in the region had returned to democracy which put an end to the repressive military dictatorships of the previous era. Hosting the Earth Summit demonstrated a new commitment to international environmental norms which was not a policy area that the military dictatorships had paid much attention to. Furthermore, it was remarkable that the summit took place in a country of the global South. Since the first UN Conference on the Human Environment held in 1972 in Stockholm, global environmental politics has been marked by significant South–North divisions in relation to how environmental issues should be addressed, who should take responsibility and which issues should be given priority. While countries of the South have argued for the need to address local environmental problems and insisted that environmental protection cannot come at the expense of economic and social development, Northern countries have tended to give greater priority to environmental concerns with global repercussions (Connolly 1996;

© The Author(s) 2017
K.M. Siegel, *Regional Environmental Cooperation in South America*,
International Political Economy Series, DOI 10.1057/978-1-137-55874-9_1

Fairman 1996; Gupta 1995; Hochstetler 2012a, 961–962; Williams 2005). The focus on environment and development evident already in the official name of the Rio Summit in 1992 thus showed the influence that the countries of the South had been able to gain over the conference (Vogler 2007, 436; Williams 2005, 56). Yet, this remains a key tension which is also central to the book.

Environmental cooperation, understood as collaborative efforts across national boundaries in order to address shared ecological concerns, has received significant scholarly attention since the Stockholm Conference in 1972. Nevertheless, two important gaps remain in the literature on environmental cooperation. First, only in the last couple of years have scholars started to look at the regional level as a distinct scale for environmental cooperation. Second, most studies have focused on Northern approaches to environmental cooperation. Countries of the South have been examined mostly in relation to environmental cooperation at the global level and in particular their opposition to Northern approaches and priorities. This means we know relatively little about how environmental cooperation between countries of the South takes place, but given the North–South disagreements that have been clearly demonstrated at the global level, approaches and priorities are likely to be different in regions of the South. In particular, social and economic development remains the foremost priority, and consequently, governments have been less concerned with creating cooperative arrangements, focussing primarily on environmental concerns.

The Southern Cone of South America presents a fascinating puzzle to study environmental cooperation in a region of the South due to the combination of two aspects. On the one hand, regional environmental cooperation in the Southern Cone has increased in quantity and quality since the early 1990s. Yet, on the other hand, regional environmental cooperation in the Southern Cone takes place in the margins of other cooperation efforts and political priorities. This makes it less visible than examples of cooperation which focus primarily on environmental concerns, and it has not received much attention in previous studies. The book therefore seeks to understand not only how robust forms of regional environmental cooperation have increasingly developed over the last two decades, but also why these have consistently remained marginalised. The aim of this analysis is twofold. Empirically, it seeks to understand the dynamics that have been driving regional environmental cooperation in the Southern Cone over the past two decades while keeping it in the margins of political agendas. Theoretically, the book aims to make a contribution to the study of environmental cooperation, particularly in regions of the South, by setting out a broad pattern of what cooperation consists of and the process that has led to cooperation in the Southern Cone. This provides a basis for comparisons with other regions. In order to do this, the book seeks to answer the following research questions: What does regional environmental cooperation in the Southern Cone consist of? Who has promoted it and how, or what has been the process leading to regional environmental cooperation? What are important characteristics?

The book examines regional environmental cooperation in the Southern Cone over a time period of about 25 years from the early 1990s onwards. It primarily focuses on how governments in the Southern Cone region cooperate on shared environmental issues and does not examine exclusively private governance arrangements. However, as the book will demonstrate very clearly, environmental cooperation between Southern Cone governments is driven by a variety of actors, many of whom are not part of the Southern Cone governments. These include domestic NGOs and networks of researchers as well as a range of different donors, including international NGOs and international organisations. In particular, the regular meetings, exchanges of information and joint projects that make up cooperation in practice are often driven by these different non-state actors. Moreover, they also play a role in encouraging and supporting governments in developing the written agreements and declarations which make up formal cooperation. While cooperation between governments is thus the point of departure of the book, the way in which cooperation works in practice means that the analysis is by no means restricted to the activities of governments.

The book concentrates on the core Southern Cone countries, namely Argentina, Brazil, Paraguay and Uruguay. Although with the creation of the new regional organisation Unasur (Unión de Naciones Suramericanas or Union of South American Nations) cooperation on other issues has over the last decade moved towards a larger scale including the whole of South America, robust environmental cooperation mostly takes place at the level of the different regions within South America. In many respects, the Southern Cone contrasts starkly with the neighbouring Amazon region which has received much more public and scholarly attention (Hochstetler and Keck 2007, 140–185; Keck 1998; Keck and Sikkink 1998, 121–163; Nepstad et al. 2009). Similar to the Amazon region, the Southern Cone is in many ways defined by a river basin. The La Plata basin, which is the second biggest basin in South America after the Amazon, links the riparian countries physically by providing an important means of transportation and a shared source of energy, exploited in a number of national and bi-national hydropower stations. Yet, the dynamics driving regional environmental cooperation are very different. Domestically, the Amazon represents the most remote areas in each of the Amazon countries, while the La Plata basin constitutes the political and economic centres of its riparian countries. Internationally, the reverse is the case, and the protection of the Amazon, especially in relation to deforestation, has been subject to considerable international pressure while the La Plata basin has received much less international attention. At the same time, the boundaries of a region are never completely fixed (Balsiger et al. 2012, 7–8; Debarbieux 2012; Elliott and Breslin 2011, 5, 13). Moreover, in environmental cooperation, the spatial boundaries are defined not only by political and economic concerns, but ecological criteria often play an important role as well. In addition, of course, developments in neighbouring countries also influence a particular region. Consequently, where relevant the book also examines some neighbouring countries and the South American context as a whole in order to situate regional environmental cooperation in the Southern Cone better.

How South American countries deal with shared environmental concerns is relevant also beyond the region. The book clearly uncovers the many political and economic links with other parts of the world that have shaped regional environmental cooperation in the Southern Cone. This means that cooperation has also been shaped by politics and choices made outside the region. Conversely, how South America's significant natural resources and energy reserves are managed also has important implications for the global climate. In some parts of the region, innovative solutions have been developed for example in relation to urban transport or forestry protection, but important challenges still remain, particularly in terms of implementing commitments made by governments and achieving social development as well as sustainability. According to Edwards and Roberts (2015, 167–169), this makes Latin American countries a bellwether for how development needs can be reconciled with sustainability.

The analysis is based on two case studies of regional environmental cooperation in the Southern Cone. In the first case, regional environmental cooperation is linked to the La Plata basin regime, a regional resource regime[1] dedicated to the sustainable development of the basin, including economic as well as environmental objectives. In this case, the regime's multilateral coordinating committee is the coordinating organisation and membership is defined by ecological as well as political and economic criteria with all five riparian states being members. The scope of issues covered is relatively broad, but always linked to the topic of freshwater resources. The second case study examines the Convention on Migratory Species (CMS), a global environmental convention which has gradually become a framework for regional environmental cooperation. Since 2005, four memoranda of understanding under the umbrella of the CMS have been signed by Southern Cone countries and some neighbouring countries. In this case, the scope of issues covered is narrower focussing specifically on endangered species that migrate across national boundaries on a regular basis and their habitats. Membership is determined by political and ecological criteria. The two case studies are contrasted with the regional organisation Mercosur, which initially seemed like a promising framework for regional environmental cooperation, but whose relevance declined over the time period examined in the book.

The case studies were selected on the basis of three different sources. First, the existing literature strongly pointed to the regional organisation Mercosur which had been the focus of most of the previous studies on regional environmental cooperation in the Southern Cone (Devia 1998; Hochstetler 2003, 2005; Laciar 2003; Tussie and Vásquez 2000). Second, from interviews and fieldwork conducted during a pilot study in 2010, it became clear that water and in particular transboundary rivers and underground water reserves are a very important environmental concern for the region. This is addressed through a range of instruments including bilateral and multilateral treaties, technical commissions and development cooperation projects that are loosely integrated under the umbrella of the La Plata basin regime. However, the regional organisation Mercosur only plays a very minor role in relation to this topic. Clearly then, Mercosur is neither the only nor the most important channel for regional environmental cooperation in the Southern Cone. Third, in order to ensure a systematic case selection, I put together an overview of treaties with clear environmental components signed between South American countries between 1940 and 2008. This overview was developed on the basis of the ECOLEX database (FAO, IUCN, and UNEP 2014), a comprehensive online database bringing together the environmental law information held by the UN Food and Agriculture Organisation (FAO), the UN Environment Programme (UNEP) and the International Union for

Conservation of Nature (IUCN). An analysis of this overview confirmed the findings from the pilot study because issues relating to water have frequently been the object of treaties between Southern Cone countries. In addition, the overview also showed that a number of agreements were made relating to global environmental conventions, an issue that did not emerge as particularly prominent either from the existing literature or from the pilot study. In particular, the Convention on Migratory Species stands out with several agreements having been signed between South American countries since 2006 within the framework of this global environmental convention. Following the case study selection, I conducted a second extended period of fieldwork between February and July 2011 in order to gain more detailed information on the three cases. My main methods of data collection were semi-structured elite interviews with policy-makers, NGOs and researchers. Altogether I conducted formal interviews with more than 50 people and several informal discussions in Argentina, Brazil, Paraguay and Uruguay as well as one interview in Germany at the CMS Secretariat[2]. In addition to the information gained directly from the interviews, many interviewees also provided me with documents or infor-mation of where to find important written documentation on the three case studies. Consequently, I also examined over 150 written articles, including reports of NGOs, governments and international organisations as well as websites, newspaper articles and research reports. I analysed this data building on grounded theory methodology, and from this, the con-cepts and processes of regional environmental cooperation set out later in the chapter emerged.

The remainder of this chapter first provides an overview of the existing literature on environmental cooperation. This uncovers two important gaps; first, a lack of studies that differentiate clearly between the global and the regional level as two distinct scales of cooperation; and second, the need for more empirical research on regional environmental cooperation and environmental politics more generally outside the global North. This is followed by an introduction of the main developments in regional envi-ronmental cooperation in the Southern Cone since the early 1990s. Based on the analysis of the Southern Cone case studies, the subsequent section outlines the different elements of cooperation and the process that has led to cooperation with the aim of providing a basis for comparison with other regions. The final section provides an overview of the book.

The Need to Look Beyond the Global North

From the Stockholm Summit in 1972 onwards, cooperation between states in order to address shared environmental concerns expanded in terms of the number of countries involved as well as the types of activity and the environmental issues covered. The academic community followed with numerous analyses of this new phenomenon of environmental cooperation, and by the 1990s, this had become a well-researched topic (Zürn 1998). However, two important caveats remain. First, until relatively recently most studies have not distinguished between the global and the regional dimension of environmental cooperation. This meant that distinct scales of cooperation that are analytically quite different have been lumped together (Balsiger and VanDeveer 2012, 7; Balsiger et al. 2012, 5). As a result, even studies of environmental cooperation on issues that are clearly much more regional rather than global in scope, such as cooperation on the Rhine basin (Bernauer 1996), the Baltic and North Seas (Haas 1993) or the Great Lakes (Valiante et al. 1997), have usually not analysed these as a form of cooperation that is analytically distinct from global environmental cooperation. The main exception to this is the European Union (EU) where environmental cooperation has been researched specifically from a regional perspective (see for example Jordan and Adelle 2013; McCormick 2001; Vogler 2011; Weale et al. 2000). The EU is generally regarded as a successful example of regional environmental cooperation and is often seen as a role model for other regions (Elliott and Breslin 2011, 1). However, the EU represents a very specific model of regional cooperation with its own dynamics which has not been replicated elsewhere. Consequently, as the book will show, it is not always helpful to apply findings from regional environmental cooperation in the EU to other regions or use the EU as the main reference point.

The lack of distinction between "global" and "regional" environmental cooperation has also made it more difficult to realise that in fact most of the research on environmental cooperation has been carried out in relation to the global North. The study of environmental cooperation has mostly focused on those cases where new institutional frameworks, or regimes, were created with the objective of addressing a particular environmental concern. Defined by Krasner (1982, 186) as "sets of implicit or explicit principles, norms, rules and decision-making procedures around which actors' expectations converge in a given area of international relations", the concept of cooperation "regime" became a valuable tool because it was able to capture many different elements of cooperation. The regime concept thus relates to official organisations, but at the same time also allows examining the less formal elements of cooperation which had not received much attention in prior studies. Regimes then are not official organisations as such, but organisations are often key players in creating and implementing regimes (Young 1999, 21). However, regime analysis has focussed on a relatively narrow set of cases of environmental cooperation because it has concentrated on new regimes that were created specifically to deal with a particular environmental issue. Even if the regime concept is no longer such a central analytical tool in the field of global environmental politics, as Clapp and Helleiner (2012, 489) point out, most studies still focus on the creation, institutional design and effectiveness of institutions, agreements and initiatives that focus explicitly on the environment. One problem that has resulted from this is that cooperation arrangements that do not focus specifically on the environment have been overlooked despite the fact that many of them have major implications for the environment. In addition, the formation of a new regime reflects a relatively high level of government commitment, and most of the cases studied demonstrate fairly high levels of institutionalisation. Research on environmental regimes has therefore left out less visible and less institutionalised cases where no new regimes have been created and where the level of government commitment is lower. Moreover, global environmental regimes have mostly been promoted by Northern countries. The UN Convention to Combat Desertification (UNCCD) is an exception in this regard as it was driven by countries of the South (Najam 2004), and with regard to climate change, some emerging powers, including Brazil, have also taken active roles in particular since 2009, by specifying voluntary domestic commitments (Hochstetler 2012a; Hochstetler and Viola 2012; Viola and Franchini 2012). Nevertheless, generally global environmental conventions have tended to be initiated and driven by countries of the North. This means that overall the study of environmental cooperation presents a Northern bias, in terms of where empirical research has been conducted geographically and in terms of the characteristics of cooperation that have been studied.

At the same time, we also know that when it comes to global environmental cooperation, priorities and approaches frequently diverge between countries of the North and South. Evidently "the global South", just like "the global North", is a huge and contested category encompassing a large variety of different countries, contexts and experiences.

Nevertheless, for a long time the G77 and China have demonstrated a high degree of unity in global environmental politics resulting from a shared understanding that the industrialised countries of the North should take responsibility for causing the deterioration of the global environment, and common interests in terms of prioritising development and insisting on additional financial resources, technology transfers and resources for capacity-building from Northern industrialised countries (Edwards and Roberts 2015, 41–44; Williams 2005). Since 2009, the emergence of new coalitions at the climate change negotiations has provided some indications that the North South division may be in the process of breaking up (Edwards and Roberts 2015, 48–51; Hochstetler 2012a; Hochstetler and Viola 2012), but the evolution of global environmental politics makes it clear that findings from studies on regional environmental cooperation in the North and Europe in particular cannot simply be applied to other regions. Clearly, the political, economic and social contexts are very different in different regions, and this means that the preconditions for regional environmental cooperation are not the same. There are of course numerous studies on regional cooperation more generally, and some of these have also specifically looked at regions outside the global North. These have found for example that regional cooperation is a way of improving domestic economies and increasing competitiveness in relation to other economic blocs (Mattli 1999, 155) or that it is a strategy of states and/or domestic actors to shape the development and impact of globalisation in their region (Grugel and Hout 1999). However, studies on regional cooperation have mostly focussed on security and economic cooperation, but left out environmental concerns (Balsiger and VanDeveer 2012, 3; Balsiger et al. 2012, 6).

At the same time, research on environmental policy-making at the national level is also more limited in relation to countries of the South compared to the North. Already in the early 1990s, Levy et al. (1993, 419) noted the need for studies on environmental policy-making in countries of the South to assess to what extent international environmental institutions make a difference. Twenty years later, other authors have pointed out that there are still major research gaps in relation to how developing and post-communist countries approach environmental protection and sustainable development and that we cannot assume that this is similar to industrialised democracies (Meadowcroft 2012, 80–81). Moreover, with regard to the climate change regime, for example, studies of the national level have mostly focussed on industrialised countries in the North. Even the increasingly important group of emerging countries has mostly been examined in relation to international negotiations, and we know very little about domestic policy processes in these countries in relation to climate change (Bailey and Compston 2012, 205–206). Scholars have also noted that, presumably because there are no international cooperation regimes, there has been relatively little research on how the rapid industrial growth in emerging economies such as China or India has impacted on global ecosystems, land use and natural resources (Clapp and Helleiner 2012, 495). Studies on climate change in Latin America have mostly focussed on technical and scientific data with few studies examining the politics, decision-making processes and policy outcomes (Edwards and Roberts 2015, 182). Finally, the activities and strategies for environmental protection of highly visible Northern actors working in countries of the South have been examined in much more detail than dynamics of national policy-making and the role of domestic actors (Steinberg 2001, 5–6).

The second important gap in the literature on environmental cooperation is thus that there is very little research on how countries in the South cooperate on shared environmental concerns and what the characteristics of cooperation are. With the publication of two important collections (Balsiger and VanDeveer 2012; Elliott and Breslin 2011) focussing specifically on regional-level environmental cooperation and including case studies from around the world, a few scholars have in the last few years started to address these gaps. However, detailed studies and theoretical tools to compare cases where regional environmental cooperation is less institutionalised are still scarce. Studying the regional level as a site of environmental cooperation is not only necessary to develop more refined theoretical accounts of environmental cooperation, but may also be useful from a practical point of view (Siegel 2016, 714–715). As there are fewer countries that have to coordinate, cooperation at the regional level may be more practical or feasible. The regional level may also be perceived as more legitimate because countries are more likely to share a common history, culture or language and North South differences which have often led to severe criticism and deadlocks of environmental cooperation at the global level are likely to be less pronounced at the regional level (Balsiger and VanDeveer 2012, 3; Elliott and Breslin 2011, 8–10). At the same time, cooperation at the regional level may also be able to act as a building block towards cooperation at the global level (Balsiger and VanDeveer 2012; Carrapatoso 2012) and help in addressing implementation gaps of global conventions (Selin 2012). This means that the regional level has the potential to complement the global level which is particularly important at a time where the global level is often perceived to be stagnating or failing to address shared environmental concerns although, of course, numerous problems also exist in regional cooperation (Balsiger and VanDeveer 2012, 2; Conca 2012).

REGIONAL ENVIRONMENTAL COOPERATION IN THE SOUTHERN CONE OF SOUTH AMERICA

Although individual initiatives also existed before, regional environmental cooperation in the Southern Cone of South America has clearly increased in quality and quantity from the early 1990s onwards. The process of democratisation that started in all the Southern Cone countries in the 1980s was crucial for regional environmental cooperation because it opened up political agendas for the inclusion of new issues, including environmental concerns, and it provided significantly more space for civil society activity. These developments facilitated the work of environmental groups as well as cross-border interaction, and the Rio Summit in 1992 further promoted the strengthening of regional environmental networks. Moreover, in an effort to gain international recognition, the new democratic states became more open towards international norms and processes. This included environmental norms as well as an increase in cooperation with external donors on environmental concerns (Hochstetler 2003, 2005, 353–356, 2012b; Mumme and Korzetz 1997). Democratisation also set the stage for enhanced cooperation between states in the region in general (Kaltenthaler and Mora 2002; Tussie and Vásquez 2000). Following a rapprochement between Argentina and Brazil, in 1991 those two countries as well as Paraguay and Uruguay signed the Treaty of Asuncion, establishing the regional organisation Mercosur. Created at a time when the Southern Cone governments adopted the neoliberal policies of the Washington Consensus, Mercosur was an example of "open regionalism" focussing on market opening and free trade. At its inception, Mercosur was thus mainly a trade agreement focussing on economic objectives as its name Mercado Común del Súr (Common Market of the South) also indicates (Carranza 2003, 68; Laciar 2003, 25; Riggirozzi 2012, 429; Torres and Diaz 2011, 203). Nevertheless, environmental concerns have also been on Mercosur's agenda since its creation with the Treaty of

Asuncion referring to the preservation of the environment in its preamble (Hochstetler 2003, 5–6, 2005, 351). Over the following years, the Mercosur member states created various institutions dedicated to environmental concerns. At its inception, Mercosur therefore had a relatively strong potential for addressing regional environmental concerns and consequently the regional organisation became the focus of much of the research on environmental cooperation in the Southern Cone (Devia 1998; Hochstetler 2003, 2005; Laciar 2003; Tussie and Vásquez 2000). Moreover, Mercosur has frequently been compared to the EU, and at least initially both, EU and Mercosur policy-makers, often expressed the view that Mercosur should follow the EU model in terms of institutional set-up (Malamud 2005, 429) or that the EU could provide a "road map" for Mercosur's institutional development (Sanchez Bajo 1999, 938). Yet, two and half decades later it has become clear that there is little political will to strengthen Mercosur and follow the EU's path of building supranational institutions. Moreover, "spillover" from economic integration to other policy areas has remained much more limited than in the European case. In this context, Mercosur's competences to address shared environmental concerns have consistently been getting weaker over time, and diverging priorities with the mostly European donors have become evident. Mercosur's environmental agenda has thus included many different topics at different points in time, but the regional organisation was not given the mandate to address several of the region's most important environmental concerns. Moreover, although several civil society networks refer to Mercosur in name and thus build on a common Mercosur identity, in fact there is little coordination with Mercosur's formal institutions.

Yet, regional environmental cooperation also started to develop in other contexts albeit less visibly and receiving less public and scholarly attention. First, since the 1990s, the La Plata basin regime has increasingly incorporated environmental concerns. Already in the late 1960s, the five riparian countries—Argentina, Bolivia, Brazil, Paraguay and Uruguay—established a regional resource regime in order to promote the economic development of the basin and in particular hydropower, while ensuring political stability. From the 1990s onwards, it was complemented with several treaties which specifically refer to environmental concerns as well as six large environmental projects with external funding which, except for one, all involved at least two countries. Second, regional networks of conservation NGOs, researchers and government officials became increasingly aware that some species were endangered and that the same species regularly migrate across national boundaries. This meant that in order to protect those species, cross-border cooperation was necessary. Consequently, regional conservation networks became increasingly active in joint monitoring as well as information exchanges and coordination of conservation measures. This cooperation was formalised in two stages. First, Argentina, Paraguay and Uruguay all joined the convention at different points in the 1990s. Formal cooperation under the CMS umbrella was strengthened further in a second phase during the 2000s when different Southern Cone and some neighbouring countries signed four memoranda of understanding outlining their commitment to the protection of different groups of species whose habitat stretches across several countries in the region. The case studies demonstrate that regional environmental cooperation in the Southern Cone is driven by a combination of endogenous and exogenous drivers.

However, the particular circumstances in which the democratic transition took place in the Southern Cone limited the extent to which regional environmental cooperation could develop. The 1980s and 1990s were thus shaped not only by the return to democracy, but also by far-reaching neoliberal reforms which made economic growth the main objective with little attention paid to other concerns. Key elements of the economic reforms carried out included measures to open the markets and reduce the role of the state as well as attracting foreign investment (Gwynne and Kay 2000). The neoliberal agenda embraced by political and economic elites in the Southern Cone and promoted by international financial institutions and the US government paved the way for more intensive natural resource exploitation with severe socio-environmental consequences (Green 1999; Murray 1999) while the newly created environmental agencies remained weak (Barton 1999, 195; Gwynne and Silva 1999, 159–160; Mumme and Korzetz 1997, 53–54).

These developments clearly provided an unfavourable context for the creation of strong forms of regional environmental cooperation. Limited state budgets and state capacities made it difficult for governments to create new regimes specifically dedicated to regional environmental concerns while external funders and NGOs came to play prominent roles. While democratisation opened up the space for the inclusion of environmental concerns on the policy agenda, neoliberal reforms simultaneously limited how far these could advance, leading to what appears like a paradox. Although robust cases of regional environmental cooperation started to develop, these remained marginalised with a high dependence on external funding or support from NGOs, a low level of government commitment resulting in non-binding and often vague agreements and the absence of regimes created specifically to address regional environmental concerns.

The 2000s were marked by important political changes when Leftist governments were voted into office in a majority of South American countries, including all the Southern Cone countries, as part of a widespread reaction of voters against the austere neoliberal reforms of the previous two decades. Although there are important differences between the different governments, they shared a commitment to poverty reduction and social objectives and sought to achieve these by using revenues from natural resource exploitation. Reversing policies adopted under the neoliberal governments of the 1980s and 1990s, Leftist governments in the Southern Cone have to varying degrees increased, or in the case of Paraguay attempted to increase, the role of the state in the management of natural resource exploitation. Benefitting from the commodity boom, revenues from the export of commodities have become an important pillar for social programmes implemented by Leftist governments and have contributed to significant improvements with regard to poverty reduction, health care provision and education. These achievements have been an important element in the popularity of progressive governments and a source of legitimacy (Gudynas 2009, 2010a, b, c; Hogenboom 2012a, b; Hogenboom and Fernández Jilberto 2009). *Neo-extractivismo* (neo-extractivism) combining old practices of natural resource exploitation with new social policies and a stronger role of the state therefore became the dominant development strategy during the 2000s (Burchardt and Dietz 2014; Gudynas 2009). In addition, the expansion of natural resource exploitation and commodity exports also served as a tool to avoid the alienation of powerful economic elites linked to the export sector which is important for political stability.

Regional cooperation too has been shaped by these developments. As social concerns and inclusion gained in importance and high commodity prices increased the autonomy of South American governments, the nature of regional cooperation also changed. In this context, two new regional organisations, the Bolivarian Alliance for the Peoples of our America or Alba, led by Venezuela and Brazilian-led Unasur, were launched. Reflecting developments at the national level, the new regional organisations shifted the focus from economic and market priorities to social concerns, such as education, health and employment as well as other domestic and regional needs, notably energy, infrastructure and security. Even if the new regional projects are still at an early stage with a very low level of institutionalisation, they are important because they have changed the parameters of regional cooperation in important ways and created new conceptions of "what regionalism *is* and *is for*" (Riggirozzi and Tussie 2012, 6). At the same time, regional integration has become increasingly "resource-driven" (Saguier 2012) due to the prominent role that natural resource exploitation plays in relation to regional cooperation. Since the start of the millennium, physical integration has become an important cornerstone of current integration projects in a region which has long looked outward and historically had a relatively poorly developed internal infrastructure (Garzón and Schilling-Vacaflor 2012; Hochstetler 2011, 140–141). The necessity of a better transport infrastructure in order to export commodities such as soybean is one of the driving forces behind such projects (Gudynas 2008, 514; Lapitz et al. 2004, 119–123). Another important element is the development of a regional energy infrastructure in order to satisfy the demands of the growing economies and in particular Brazil (Burges 2005; Garzón and Schilling-Vacaflor 2012).

Over the past decade, the focus of governments has thus been on addressing social concerns and strengthening the autonomy of the region. Natural resource exploitation has been central for both of these objectives, and this has led to the sidelining of socio-environmental concerns. The adoption of neo-extractivism as the main development model dashed the hopes that alternative and radically different approaches to development and environmental governance would gain ground under progressive governments (Silva 2016, 327–328). At the same time, South American governments have continued to participate in and shape global environmental politics and there is a general recognition that environmental concerns cannot be ignored. As a consequence, regional environmental cooperation has continued to develop driven by domestic and external actors, but it has remained in the margins of regional cooperation and policy agendas. 2016 has seen important new political developments in Argentina where elections have once more brought a centre-right government to power, and Brazil where a controversial impeachment process on highly dubious grounds has brought down the government of Dilma Rousseff of the centre-left Worker's Party. This shift to the right is likely to entail different priorities in regional cooperation, but it is unlikely that environmental concerns will gain more attention.

Overall, the apparent paradox of robust regional environmental cooperation in the margins is thus an outcome of the political and economic context which has shaped policy-making at the national level as well as regional cooperation in the Southern Cone in the last two decades. In particular, the development strategy adopted by governments has been built on large scale and intensive natural resource exploitation and reinforced old tensions between development and socio-environmental concerns. Nevertheless, pressure from various regional civil society groups and research networks often working together with some government officials as well as international organisations and external funders has succeeded in putting shared environmental concerns on regional political agendas in various settings. In the following section, I will outline what regional environmental cooperation in the Southern Cone consists of, what key characteristics are and what central steps in the process leading to cooperation have been. Recognising that regional environmental cooperation in regions of the South looks very different from the North and the European case in particular, the aim is to identify patterns that can be compared with other regions. This is an important starting point for further research given that most empirical case studies and theories in the social sciences are still based on OECD countries (Narlikar 2016), and this trend is also reflected in the field of environmental cooperation.

COMPONENTS, DRIVERS AND PROCESSES OF COOPERATION

Reflecting observations of the few studies that have examined other regions in the South, an important characteristic of environmental cooperation in the Southern Cone is its perceived marginality. Given the significant challenges that cooperation often faces, a central concern in the study of environmental cooperation in regions of the South is how weak or strong a particular cooperation effort is, how to compare different regions in relation to this and how to deal with or overcome marginality. In this section, I argue that it is helpful to distinguish between regular activities of environmental cooperation and formal cooperation consisting of agreements between countries. Both components can occur separately, but when they become linked in relation to the same environmental issue cooperation becomes significantly more robust. The process leading to robust cases of regional environmental cooperation linking formal agreements with regular activities is an incremental one and shaped by various actors at different stages. Initially, a demand for regional environmental cooperation emerged from an increased interest and knowledge of the transboundary dimensions of an environmental concern. Networks of researchers and civil society organisations are an important endogenous driver at this stage. However, as governments only make limited funding available, exogenous drivers in the form of external funders become a crucial actor in making cooperation in practice possible. As a result, cooperation is shaped to some extent by the objectives and priorities of external funders. In addition, governmental approval is necessary for written agreements between states, and governments decide which topics, approaches and framings proposed by endogenous or exogenous drivers are taken up in formal cooperation.

The studies of the emerging research agenda on environmental cooperation outside Europe seem to share a concern with the relative weakness of regional environmental cooperation (Siegel 2016, 721). Kulauzov and Antypas (2011, 128) describe both environmental and regional cooperation in the Middle East and North Africa as "marginal phenomena" while Compagnon, Florémont and Lamaud (2011, 106) characterise regional environmental governance in sub-Saharan Africa as "embryonic" and "fragmented". Swain (2011, 87) notes the "lack of political will" in South Asia, and Matthew (2012, 111) describes initiatives of the South Asia Association for Regional Cooperation as "modest actions with little tangible effect". Finally, Hochstetler (2011, 145) points out the "institutional fragility" of regional environmental governance arrangements in South America and Elliott (2011, 71) points to the lack of funding for environmental concerns in both Northeast and Southeast Asia. Similarly, the Southern Cone case studies examined in this book give the impression that regional environmental cooperation is not a political priority and faces a number of significant challenges. A notable characteristic of regional environmental cooperation in the Southern Cone is therefore its marginality compared to other policy areas and compared to the stronger cases of environmental cooperation that the majority of previous studies have focussed on. This marginality consists of three key elements: first, the absence of cooperation regimes created specifically to address regional environmental concerns. Instead, regional environmental cooperation takes place in the framework of regional or international institutions established previously and for other purposes. Second, although regional environmental cooperation has increasingly become formalised in written agreements between governments since the 1990s, many of these are vague which makes implementation harder, or non-binding. Third, there is a high dependence on external funding or the support of NGOs in order to make cooperation in practice happen. Taken together, these three elements define the marginality of regional environmental cooperation in the Southern Cone and demonstrate the extent to which cooperation is a low government priority. This helps to draw out differences compared to regions in the North where, even though governments also have other priorities, they tend to dedicate more resources to addressing shared environmental concerns and outline more specific objectives, and levels of institutionalisation are often higher.

A key concern that emerges from the observation that environmental cooperation is relatively weak in many regions of the South is the question of what this means for cooperation, for example how likely cooperation efforts are to endure and what they might achieve. Yet, there are few comparisons between different regions, and it is difficult to say whether cooperation is stronger, weaker or more likely to endure in one region than in another. Studies on environmental regimes have addressed the question of effectiveness (Haas et al. 1993) and thus examined different levels of strength of environmental cooperation. However, by focussing on environmental regimes, these relate to the stronger end of the spectrum of environmental cooperation where government commitment tends to be relatively high. For the lower end of the spectrum where environmental cooperation is less institutionalised and government commitment is lower, theoretical tools to assess the level of strength or weakness are much less well developed. Given how little research has been carried out on less visible and more marginal cases of regional environmental cooperation, assessing its effectiveness would be an enormous task. In this book, as a first step, I therefore focus on the question of robustness, that is the ability of cooperation to continue when circumstances change (Hasenclever et al. 1996, 178). This is conceptually different from effectiveness which relates to how successful cooperation has been in addressing a certain environmental concern.

In order to develop a better understanding of less institutionalised and more marginal cases of regional environmental cooperation, and in particularly its level of robustness, I propose to disaggregate cooperation explicitly into two elements of cooperation. The first element is what I call "formal" environmental cooperation which refers to written agreements, declarations or joint policy statements. These are negotiated at the highest levels of government, and governments are the main actors in formal cooperation. The agreements or declarations that make up formal cooperation represent a written and public commitment of states to regional environmental cooperation on a particular topic. "Commitment" can be understood as accepting a particular norm as valid and binding for themselves (Risse and Ropp 2013, 9). The desire to gain a good reputation and international legitimacy can be important factors for states to make such commitments (Finnemore and Sikkink 1998, 902–906; Risse and Ropp 2013, 20–21). Formal cooperation plays a crucial role in promoting continuity, because a formal agreement between states is a public and written commitment which remains valid even when governments or donors change. The findings from the book show that this in turn makes it easier to source funding and civil society organisations can also use formal and public commitments to hold governments to account and remind them of their obligations using what Keck and Sikkink (1998, 24) call "accountability politics". However, commitments can be stronger or weaker. In the Southern Cone cases examined in the book, non-binding agreements are more common and therefore my definition of formal cooperation includes legally binding as well as non-binding agreements. Moreover, the agreements that make up formal cooperation can be more or less specific. Overall, my definition of formal cooperation therefore includes a continuous range of written and public agreements between governments ranging from non-binding and fairly general declarations reflecting a relatively low level of commitment by governments, to binding and specific treaties representing a much stronger commitment. The second element is what I call cooperation "in practice". This refers to the daily or very regular joint or coordinated activities between partners in different countries. Cooperation in practice is important in order to make sure that cooperation also happens on the ground, for example through common monitoring of environmental concerns, implementing specific conservation measures or developing joint approaches to shared problems. Regular meetings or exchanges of information and best practices are important elements of cooperation in practice. The actors involved in cooperation in practice are state officials, typically lower-level government officials working for environmental agencies, as well as non-state actors, notably environmental NGOs. Moreover, cooperation in practice is often supported by external actors, notably different types of funders and international organisations.

Distinguishing between different aspects of cooperation is of course not a new idea. In fact, the regime concept was developed precisely in order to capture more and less formal aspects of cooperation. However, it is important to recognise that formal cooperation and cooperation in practice can take place at the same time and in relation to the same environmental issue, but that this is not necessarily the case and one element can also be decoupled from the other. Moreover, they do not signify different stages of environmental cooperation where one automatically follows on from the other. Instead, cooperation in practice and formal cooperation are two different elements of regional environmental cooperation which may be more or less linked. In the South American context of this book, the distinction between these two elements of cooperation became a crucial aspect of the analysis, which was also vital in order to select comparatively robust cases of regional environmental cooperation. During the first fieldwork period, I thus realised that cooperation in practice can and does often exist without formal cooperation and vice versa. However, with a high dependence on external funding and few specific commitments on the part of governments as was the case for the Southern Cone, environmental cooperation is very vulnerable if only one of the elements is present. On the one hand, there is always the risk that cooperation in practice may stop when funding runs out or when a particular project with an external donor comes to an end. Formal cooperation on the other hand, risks becoming an empty phrase on paper if it is not implemented due to a lack of political will or resources. As several scholars of Latin American politics have noted, formal institutions very often only tell half the story and there can be a huge gap between the rules that are set out on paper and actual practices (Arias 2009, 240; Grugel 2007, 242). One of the key concerns of the book was therefore to understand in which circumstances regional environmental cooperation becomes more robust, that is more likely to continue beyond an initial starting period. If cooperation in practice becomes linked to formal cooperation on the same environmental issue, cooperation is more likely to continue even if external circumstances change. It is thus possible for regional environmental cooperation to be robust even if no new regime has been created and it takes place in the margins of political agendas. Robustness then occupies a middle ground on a continuum between weak cooperation on the one hand where only one of the elements of cooperation may be present and which is more likely to stall, and strong cooperation on the other hand where institutions are stronger, funding is more secure and commitments of governments more specific.

The process leading to robust cases of regional environmental cooperation is one of incremental change that takes place over years or even decades. Although many of the actors promoting regional environmental cooperation in the Southern Cone are similar to those in the global North and include NGOs, researchers and lower-level government officials, due to the reliance on external funding the role of funders from outside the region is much more significant not only in making cooperation possible, but also in shaping its characteristics. Furthermore, national governments are not a main driver for regional environmental cooperation, but they do play a crucial role in determining which civil society organisations have access to regional and international institutions and which ideas and objectives are translated into the written agreements that make up formal cooperation. Overall, the emergence of robust cases of regional environmental cooperation has therefore been shaped by different actors during three key stages.

In the first instance, an increased knowledge, awareness and interest in the transboundary nature of a given environmental concern have played a crucial role. Networks of researchers based in different countries and in different institutions including universities, government agencies and NGOs have been important in fostering this understanding of the transboundary dimensions leading to the realisation that the environmental concern in question can only be successfully addressed by several countries working together. As noted also in other studies, an understanding of the transboundary nature and the extent of environmental problems drives the demand for regional cooperation among governments in order to minimise transboundary externalities or manage shared resources (Elliott and Breslin 2011, 7). However, in many countries of the South, governments lack the resources to research the complex transboundary dimensions of environmental concerns, and as a result, non-state actors are often important endogenous drivers for regional environmental cooperation. In both case, studies examined in this book, regional networks of researchers displaying many of the characteristics of "epistemic communities" (Haas 1992) have played an important role in the initial research. The members of epistemic communities do not necessarily work as scientists, but can also work in government agencies or NGOs (Cross 2013, 147, 153–154). In the Southern Cone cases this has been the case and networks of researchers have been important in linking government agencies, universities, NGOs and international organisations. Moreover, particularly in the case of the La Plata basin, civil society networks have also formed and debated social and political questions in relation to how shared water resources should be managed and what the priorities should be.

Although from the 1990s onwards the Southern Cone governments have increasingly recognised the importance of taking environmental concerns into account and this provided a much more favourable context for non-state actors to promote regional environmental cooperation, cooperation remains very much dependent on external funding. In order to make cooperation happen in practice, an important second step is therefore establishing sources of funding. As other studies have also noted, important barriers to more successful regional environmental cooperation often include insufficient resources, technology and expertise, including a limited understanding of how environmental concerns relate to other policy areas, and a lack of capacity including the capacity to implement policies, as well as, in some cases a lack of political will (Elliott and Breslin 2011, 16–17). Environmental aid has the potential to affect political dynamics positively by increasing concern and improving capabilities and the contractual environment (Connolly 1996, 362–363; Keohane 1996, 1–14). Provided that strategies are carefully designed and take into account recipients' priorities, environmental aid can help to change the incentives of key actors, strengthen domestic coalitions interested in environmental protection and enhance capacity (Connolly 1996, 328). A lack of capacity can be an important factor for states not to comply with international norms even if they are willing to do so (Risse and Ropp 2013, 15). If external funders target specifically transboundary or shared environmental concerns in a particular region, they can also strengthen regional environmental cooperation by supporting regular activities and providing incentives for formal agreements. The reliance on external funding makes international funders an important exogenous driver for regional environmental cooperation, but it also means that actors from outside the region have a significant influence in shaping the nature and characteristics of cooperation. The Southern Cone cases clearly demonstrate how external funders become involved in the process of building regional environmental cooperation and shape where and how it takes place. By prioritising certain environmental issues, geographical locations or institutional frameworks exogenous drivers can thus shape regional environmental cooperation in particular ways while cooperation on other issues which are not able to attract funding can be much harder. Obtaining funding to carry out regular cooperation activities in practice is therefore a central stage in the process leading to robust examples of regional environmental cooperation. In this respect, regional environmental cooperation in the Southern Cone differs significantly from case studies in the global North where external funders play much less of a role.

In addition to external funders, national governments also intervene and shape the process of regional environmental cooperation. Although governments have not emerged as key drivers for regional environmental cooperation in the Southern Cone, the agreements that make up formal cooperation nevertheless need governmental consent. The position of national governments is therefore important in terms of determining which topics, approaches and framings of shared or transboundary environmental concerns are taken up in formal agreements. Moreover, non-state actors often seek access to regional or international institutions in an attempt to get their concerns included on political agendas. National governments also play an important role in deciding which non-state actors gain access to regional or international institutions and what kinds of issues and framings are included on political agendas for debate and for translation into formal agreements.

On the whole, robust cases of regional environmental cooperation linking formal agreements with regular activities have thus been shaped by various actors at different stages. However, it is important to note that the stages are not automatic and they may not necessarily occur in that order. In particular, the search for funding is often a continuous process as projects end and new funding is sought for new initiatives. This means that there may well be different funders which shape cooperation at the same time or at different points in time. Moreover, there can also be interaction between the different stages in the process. The prospect of gaining external funding, for example, might increase the motivation of governments to sign a formal agreement, but conversely, a formal agreement might also make it easier to get funding.

Outline of the Book

The book is broken down into five further chapters. Chapters 2 and 3 examine in more depth the political, economic and social context and the processes of regional integration in the Southern Cone which are crucial to understand the possibilities and limitations for regional environmental cooperation. Chapter 2 seeks to explain not only how robust forms of regional environmental cooperation were able to develop over the past 25 years, but also why they have remained marginal. The chapter outlines how the return to democracy opened up political spaces and agendas and made the development of robust forms of regional environmental cooperation possible. Yet, at the same time, the neoliberal reform agenda limited how far regional environmental cooperation could advance. The marginality of socio-environmental concerns and in particular those relating to resource exploitation was further reinforced with the adoption of neo-extractivist development strategies across the region in the 2000s. Chapter 3 focuses on the evolution of regionalism in South America with the aim of setting out how regional environmental cooperation has come to develop largely separately from government-led regional integration projects. This uncovers the declining relevance over time of the regional organisation Mercosur for environmental cooperation and a more general trend of sidelining socio-environmental concerns in government-led regional organisations and integration processes. However, more robust examples of regional environmental cooperation have developed in other frameworks discussed in Chaps. 4 and 5. Chapter 4 focusses on the La Plata basin regime, a regional resource regime that has received little attention in previous studies on regional cooperation. Here regional environmental cooperation has developed following the work of regional research networks and with funding from external donors, notably the Global Environmental Facility (GEF). The case study also demonstrates the role that governments have played in closing down spaces for regional environmental cooperation and civil society participation as well as tensions over the involvement of external donors. Networks of researchers were also central in initiating and promoting regional cooperation on migratory species discussed in Chap. 5. Based on close cooperation between NGOs, researchers and government officials with the support of a variety of donors and the secretariat of the CMS, several agreements were signed between Southern Cone countries to protect migratory species. Finally, the concluding chapter discusses three key findings that emerge from the comparison of the findings presented in the previous chapters. First, the findings suggest that there are considerable differences in the influence that different drivers are able to exert on regional environmental cooperation. While this is not surprising, it is important because it shapes the characteristics of environmental cooperation and it is also a concern with respect to the legitimacy of cooperation. Second, the findings indicate that government-led regional integration processes are not a driver for regional environmental cooperation and on the contrary have on several occasions presented obstacles. This is more unexpected and it is important because it offers a better understanding of the nature of regionalism in South America and raises questions about the analytical value of the EU as a model for regional cooperation elsewhere. Third, the analysis presented in the book suggests that the marginality of environmental cooperation in the Southern Cone is closely related to both, the development strategy adopted by governments and the region's position in the global political economy. This is relevant for those seeking to strengthen regional environmental cooperation in South America, and also points to questions for further research in other regions of the South.

NOTES

1. Given the different circumstances and economic and political objectives in different regions it is not surprising that there are also very different types of regional organisations, in terms of objectives, institutional structures and level of development. This has posed challenges in terms of developing a conceptual framework and agreeing on definitions or comparing developments in different regions (Hettne 2005; Hurrell 1995, 333; De Lombaerde et al. 2010). For the purpose of the book which focusses on regional *environmental* cooperation rather than regional integration in general, I distinguish between "regional resource regimes" which were created specifically in order to manage a particular shared resource and "regional organisations" as an overall umbrella to denote those organisations that were created by states in a region to achieve vital political or economic aims which do not primarily relate to transboundary natural resources. However, both of these can contribute to regional integration in different ways.

2. The majority of interviews were conducted in Spanish, but some were in English or German. Almost all interviewees agreed to be cited including their organisational affiliation. Those who preferred to remain anonymous or off the record are not referred to in the book.

REFERENCES

Arias, Enrique Desmond. 2009. Ethnography and the Study of Latin American Politics: An Agenda for Research. In *Political Ethnography—What Immersion Contributes to the Study of Power*, ed. Edward Schatz. Chicago: University of Chicago Press.

Bailey, Ian, and Hugh Compston. 2012. Political Strategy and Climate Policy in Rapidly Industrializing Countries. In *Feeling the Heat—The Politics of Climate Policy in Rapidly Industrializing Countries*, ed. Ian Bailey and Hugh Compston. Houndmills: Palgrave Macmillan.

Balsiger, Jörg, and Stacy D. VanDeveer. 2012. Navigating Regional Environmental Governance. *Global Environmental Politics* 12 (3): 1–17.

Balsiger, Jörg, Miriam Prys, and Niko Steinhoff. 2012. *The Nature and Role of Regional Agreements in International Environmental Politics: Mapping Agreements, Outlining Future Research.* GIGA Working Paper 208. Hamburg: German Institute of Global and Area Studies. http://www.giga-hamburg.de/index.php?file=workingpapers.html&folder=publikationen.

Barton, Jonathan R. 1999. The Environmental Agenda: Accountability for Sustainability. In *Developments in Latin American Political Economy—States, Markets and Actors*, ed. Julia Buxton and Nicola Phillips. Manchester: Manchester University Press.

Bernauer, Thomas. 1996. Protecting the Rhine River against Chloride Pollution. In *Institutions for Environmental Aid*, ed. Robert O. Keohane and Marc A. Levy. London: MIT Press.

Burchardt, Hans-Jürgen, and Kristina Dietz. 2014. (Neo-)extractivism—A New Challenge for Development Theory from Latin America. *Third World Quarterly* 35 (3): 468–486.

Burges, Sean W. 2005. Bounded by the Reality of Trade: Practical Limits to a South American Region. *Cambridge Review of International Affairs* 18 (3): 437–454.

Carranza, Mario E. 2003. Can Mercosur Survive? Domestic and International Constraints on Mercosur. *Latin American Politics and Society* 45 (2): 67–103.

Carrapatoso, Astrid. 2012. In Search of Alternative Governance Models—The Contribution of Interregional Climate Cooperation to the Global Climate Change Regime. In *Earth System Governance Conference, Lund, Sweden, 18–20 April 2012.*

Clapp, Jennifer, and Eric Helleiner. 2012. International Political Economy and the Environment: Back to the Basics? *International Affairs* 88 (3): 485–501.

Compagnon, Daniel, Fanny Florémont, and Isabelle Lamaud. 2011. Sub-Saharan Africa—Fragmented Environmental Governance without Regional Integration. In *Comparative Environmental Regionalism*, ed. Lorraine Elliott and Shaun Breslin. Oxon: Routledge.

Conca, Ken. 2012. The Rise of the Region in Global Environmental Politics. *Global Environmental Politics* 12 (3): 127–133.

Connolly, Barbara. 1996. Increments for the Earth: The Politics of Environmental Aid. In *Institutions for Environmental Aid*, ed. Robert O. Keohane and Marc A. Levy. London: MIT Press.

Cross, Mai'a K. Davis. 2013. Rethinking Epistemic Communities Twenty Years Later. *Review of International Studies* 39 (01): 137–160.

Debarbieux, Bernard. 2012. Commentary—How Regional Is Regional Environmental Governance? *Global Environmental Politics* 12 (3): 119–126.

Devia, Leila (ed.). 1998. *Mercosur Y Medio Ambiente*, 2a ed. Buenos Aires: Ediciones Ciudad Argentina.

Edwards, Guy, and J. Timmons Roberts. 2015. *A Fragmented Continent—Latin America and the Global Politics of Climate Change*. Cambridge, MA: MIT Press.

Elliott, Lorraine. 2011. East Asia and Sub-Regional Diversity—Initiatives, Institutions and Identity. In *Comparative Environmental Regionalism*, ed. Lorraine Elliott and Shaun Breslin. Oxon: Routledge.

Elliott, Lorraine, and Shaun Breslin. 2011. Researching Comparative Regional Environmental Governance—Causes, Cases and Consequences. In *Comparative Environmental Regionalism*, ed. Lorraine Elliott and Shaun Breslin. Oxon: Routledge.

Fairman, David. 1996. The Global Environment Facility: Haunted by the Shadow of the Future. In *Institutions for Environmental Aid*, ed. Robert O. Keohane and Marc A. Levy. London: MIT Press.

FAO, IUCN, and UNEP. 2014. Ecolex. http://www.ecolex.org.

Finnemore, Martha, and Kathryn Sikkink. 1998. International Norm Dynamics and Political Change. *International Organization* 52 (4): 887–917.

Garzón, Jorge, and Almut Schilling-Vacaflor. 2012. *Infrastrukturprojekte Zwischen Geopolitischen Interessen Und Lokalen Konflikten*. 10/2012. GIGA Focus Lateinamerika. Hamburg: German Institute of Global and Area Studies. http://www.giga-hamburg.de/de/publikationen/giga-focus/lateinamerika.

Green, Duncan. 1999. A Trip to the Market: The Impact of Neoliberalism in Latin America. In *Developments in Latin American Political Economy—States, Markets and Actors*, ed. Julia Buxton and Nicola Phillips. Manchester: Manchester University Press.

Grugel, Jean. 2007. Latin America after the Third Wave. *Government and Opposition* 42 (2): 242–249.

Grugel, Jean, and Wil Hout (eds.). 1999. *Regionalism across the North-South Divide*. London: Routledge.

Gudynas, Eduardo. 2008. The New Bonfire of Vanities: Soybean Cultivation and Globalization in South America. *Development* 51 (4): 512–518.

Gudynas, Eduardo. 2009. Diez Tesis Urgentes Sobre El Nuevo Extractivismo. Contextos Y Demandas Bajo El Progresismo Sudamericano Actual. In *Extractivismo, Política Y Sociedad*. Quito: CAAP (Centro Andino de Acción Popular) and CLAES (Centro Latino Americano de Ecología Social).

Gudynas, Eduardo. 2010a. La Ecología Política Del Progresismo Sudamericano: Los Límites Del Progreso Y La Renovacíon Verde de La Izquierda. *Sin Permiso* (8): 147–167.

Gudynas, Eduardo. 2010b. Si Eres Tan Progresista ¿Por Qué Destruyes La Naturaleza? Neoextractivismo, Izquierda Y Alternativas. *Ecuador Debate* (79): 61–81.

Gudynas, Eduardo. 2010c. Agropecuaria Y Nuevo Extractivismo Bajo Los Gobiernos Progresistas de América Del Sur. *Territorios* (5): 37–54.

Gupta, Joyeeta. 1995. The Global Environment Facility in Its North-South Context. *Environmental Politics* 4 (1): 19–43.

Gwynne, Robert N., and Cristobal Kay. 2000. Views from the Periphery: Futures of Neoliberalism in Latin America. *Third World Quarterly* 21 (1): 141–156.

Gwynne, Robert N., and Eduardo Silva. 1999. The Political Economy of Sustainable Development. In *Latin America Transformed—Globalization and Modernity*, ed. Robert N. Gwynne and Cristobal Kay. London: Arnold.

Haas, Peter M. 1992. Introduction: Epistemic Communities and International Policy Coordination. *International Organization* 46 (1): 1–35.

Haas, Peter M. 1993. Protecting the Baltic and North Seas. In *Institutions for the Earth—Sources for Effective Environmental Protection*, ed. Peter M. Haas, Robert O. Keohane, and Marc A. Levy. London: MIT Press.

Haas, Peter M., Robert O. Keohane, and Marc A. Levy (eds.). 1993. *Institutions for the Earth—Sources for Effective Environmental Protection*. London: MIT Press.

Hasenclever, Andreas, Peter Mayer, and Volker Rittberger. 1996. Interests, Power, Knowledge: The Study of International Regimes. *Mershon International Studies Review* 40: 177–228.

Hettne, Björn. 2005. Beyond the 'New' Regionalism. *New Political Economy* 10 (4): 543–571.

Hochstetler, Kathryn. 2003. Fading Green? Environmental Politics in the Mercosur Free Trade Agreement. *Latin American Politics and Society* 45 (4): 1–32.

Hochstetler, Kathryn. 2005. Race to the Middle: Environmental Politics in the Mercosur Free Trade Agreement. In *Handbook of Global Environmental Politics*, ed. Peter Dauvergne. Cheltenham: Edward Elgar.

Hochstetler, Kathryn. 2011. Under Construction—Debating the Region in South America. In *Comparative Environmental Regionalism*, ed. Lorraine Elliott and Shaun Breslin. Oxon: Routledge.

Hochstetler, Kathryn. 2012a. Climate Rights and Obligations for Emerging States: The Cases of Brazil and South Africa. *Social Research* 79 (4): 957–982.

Hochstetler, Kathryn. 2012b. Democracy and the Environment in Latin America and Eastern Europe. In *Comparative Environmental Politics—Theory, Practice, and Prospects*, ed. Paul F. Steinberg and Stacy D. VanDeveer. London: MIT Press.

Hochstetler, Kathryn, and Margaret E. Keck. 2007. *Greening Brazil—Environmental Activism in State and Society*. Durham: Duke University Press.

Hochstetler, Kathryn, and Eduardo Viola. 2012. Brazil and the Politics of Climate Change: Beyond the Global Commons. *Environmental Politics* 21(5): 753–771.

Hogenboom, Barbara. 2012a. The Return of the State and New Extractivism: What about Civil Society? In *Civil Society and the State in Left-Led Latin America—Challenges and Limitations to Democratization*, ed. Barry Cannon and Peadar Kirby. London: Zed Books.

Hogenboom, Barbara. 2012b. Depoliticized and Repoliticized Minerals in Latin America. *Journal of Developing Societies* 28 (2): 133–158.

Hogenboom, Barbara, and Alex E Fernández Jilberto. 2009. The New Left and Mineral Politics: What's New? *European Review of Latin American and Caribbean Studies* (87): 93–102.

Hurrell, Andrew. 1995. Explaining the Resurgence of Regionalism in World Politics. *Review of International Studies* 21 (October): 331–358.

Jordan, Andrew, and Camilla Adelle (eds.). 2013. *Environmental Policy in the EU —Actors, Institutions and Processes*, 3rd ed. Oxon: Routledge.

Kaltenthaler, Karl, and Frank O. Mora. 2002. Explaining Latin American Economic Integration: The Case of Mercosur. *Review of International Political Economy* 9 (1): 72–97.

Keck, Margaret E. 1998. Planafloro in Rondônia: The Limits of Leverage. In *The Struggle for Accountability—The World Bank, NGOs and Grassroots Movements*, ed. Jonathan A. Fox and L. David Brown. London: MIT Press.

Keck, Margaret E., and Kathryn Sikkink. 1998. *Activists Beyond Borders—Advocacy Networks in International Politics*. London: Cornell University Press.

Keohane, Robert O. 1996. Analyzing the Effectiveness of International Environmental Institutions. In *Institutions for Environmental Aid*. London: MIT Press.

Krasner, Stephen D. 1982. Structural Causes and Regime Consequences: Regimes as Intervening Variables. *International Organization* 36 (02): 185–205.

Kulauzov, Dora, and Alexios Antypas. 2011. The Middle East and North Africa— Sub-Regional Environmental Cooperation as a Security Issue. In *Comparative Environmental Regionalism*, ed. Lorraine Elliott and Shaun Breslin. Oxon: Routledge.

Laciar, Mirta Elizabeth. 2003. *Medio Ambiente Y Desarrollo Sustentable*. Buenos Aires: Editorial Ciudad Argentina.

Lapitz, Rocío, Gerardo Evia, and Eduardo Gudynas. 2004. *Soja Y Carne En El Mercosur - Comercio, Ambiente Y Desarrollo Agropecuario*. Montevideo: Coscoroba Ediciones.

Levy, Marc, Robert O. Keohane, and Peter M Haas. 1993. Improving the Effectiveness of International Environmental Institutions. In *Institutions for the Earth—Sources for Effective Environmental Protection*, ed. Peter M. Haas, Robert O. Keohane, and Marc A. Levy. London: MIT Press.

Lombaerde, De, Fredrik Söderbaum Philippe, Luk Van Langenhove, and Francis Baert. 2010. The Problem of Comparison in Comparative Regionalism. *Review of International Studies* 36 (03): 731–753.

Malamud, Andrés. 2005. Mercosur Turns 15: Between Rising Rhetoric and Declining Achievement. *Cambridge Review of International Affairs* 18 (3): 421–436.

Matthew, Richard. 2012. Environmental Change, Human Security, and Regional Governance: The Case of the Hindu Kush/Himalaya Region. *Global Environmental Politics* 12 (3): 100–118.

Mattli, Walter. 1999. *The Logic of Regional Integration—Europe and Beyond*. Cambridge: Cambridge University Press.

McCormick, John. 2001. *Environmental Policy in the European Union*. Houndmills: Palgrave.

Meadowcroft, James. 2012. Greening the State? In *Comparative Environmental Politics—Theory, Practice, and Prospects*, ed. Paul F. Steinberg and Stacy D. VanDeveer. London: MIT Press.

Mumme, Stephen P., and Edward Korzetz. 1997. Democratization, Politics, and Environmental Reform. In *Latin American Environmental Policy in International Perspective*, ed. Gordon J. MacDonald, Daniel L. Nielson, and Marc A. Stern. Oxford: Westview Press.

Murray, Warwick E. 1999. Natural Resources, the Global Economy and Sustainability. In *Latin America Transformed—Globalization and Modernity*, ed. Robert N. Gwynne and Cristobal Kay. London: Arnold.

Najam, Adil. 2004. Dynamics of the Southern Collective: Developing Countries in Desertification Negotiations. *Global Environmental Politics* 4 (3): 128–154.

Narlikar, Amrita. 2016. *Weil Sie Wichtig Sind. Vielfalt Anerkennen, Forschung Globalisieren*. 1/2016. GIGA Focus Global. Hamburg: German Institute of Global and Area Studies. https://www.giga-hamburg.de/de/publikationen/giga-focus/global.

Nepstad, Daniel, Britaldo S. Soares-Filho, Frank Merry, André Lima, Paulo Moutinho, John Carter, Maria Bowman, Andrea Cattaneo, Hermann Rodrigues, Stephan Schwartzman, David G. McGrath, Claudia M. Stickler, Ruben Lubowski, Pedro Piris-Cabezas, Sergio Rivero, Ane Alencar, Oriana Almeida and Osvaldo Stella. 2009. The End of Deforestation in the Brazilian Amazon. *Science* 326: 1350–1351.

Riggirozzi, Pía. 2012. Region, Regionness and Regionalism in Latin America: Towards a New Synthesis. *New Political Economy* 17 (4): 421–443.

Riggirozzi, Pía, and Diana Tussie. 2012. The Rise of Post-Hegemonic Regionalism in Latin America. In *The Rise of Post-Hegemonic Regionalism—The Case of Latin America*, ed. Pía Riggirozzi and Diana Tussie. Dordrecht: Springer.

Risse, Thomas, and Stephen C. Ropp. 2013. Introduction and Overview. In *The Persistent Power of Human Rights—From Commitment to Compliance*, ed. Thomas Risse, Stephen C. Ropp, and Kathryn Sikkink. Cambridge: Cambridge University Press.

Saguier, Marcelo. 2012. Socio-Environmental Regionalism in South America: Tensions in New Development Models. In *The Rise of Post-Hegemonic Regionalism—The Case of Latin America*, ed. Pía Riggirozzi and Diana Tussie. Dordrecht: Springer.

Sanchez Bajo, Claudia. 1999. The European Union and Mercosur: A Case of Inter-Regionalism. *Third World Quarterly* 20 (5): 927–941.

Selin, Henrik. 2012. Global Environmental Governance and Regional Centers. *Global Environmental Politics* 12 (3): 18–37.

Siegel, Karen M. 2016. Can Regional Cooperation Promote Sustainable Development? In *The Palgrave Handbook of International Development*, ed. Jean Grugel and Daniel Hammett. Basingstoke: Palgrave Macmillan.

Silva, Eduardo. 2016. Afterword: From Sustainable Development to Environmental Governance. In *Environmental Governance in Latin America*, ed. Fabio de Castro, Barbara Hogenboom, and Michiel Baud. Basingstoke: Palgrave Macmillan.

Steinberg, Paul F. 2001. *Environmental Leadership in Developing Countries—Transnational Relations and Biodiversity Policy in Costa Rica and Bolivia*. London: MIT Press.

Swain, Ashok. 2011. South Asia, Its Environment and Regional Institutions. In *Comparative Environmental Regionalism*, ed. Lorraine Elliott and Shaun Breslin. Oxon: Routledge.

Torres, Alicia, and José Pedro Diaz. 2011. MERCOSUR Ambiental: ¿se Trata de Una Mirada Sólo Desde El Comercio O Del Avance de La Dimensión Olvidada? ¿Medio Lleno O Medio Vacío? In *Mercosur 20 Años*, ed. Gerardo Caetano. Montevideo: CEFIR. http://cefir.org.uy/documentacion/publicaciones-cefir.

Tussie, Diana, and Patricia Vásquez. 2000. Regional Integration and Building Blocks: The Case of Mercosur. In *The Environment and International Trade Negotiations—Developing Country Stakes*, ed. Diana Tussie. Houndmills: Macmillan Press Ltd.

Valiante, Marcia, Paul Muldoon, and Lee Botts. 1997. Ecosystem Governance: Lessons from the Great Lakes. In *Global Governance—Drawing Insights from the Environmental Experience*, ed. Oran R. Young. London: MIT Press.

Viola, Eduardo, and Matias Franchini. 2012. Climate Politics in Brazil: Public Awareness, Social Transformations and Emissions Reductions. In *Feeling the Heat—The Politics of Climate Policy in Rapidly Industrializing Countries*, ed. Ian Bailey and Hugh Compston. Houndmills: Palgrave Macmillan.

Vogler, John. 2007. The International Politics of Sustainable Development. In *Handbook of Sustainable Development*, ed. Giles Atkinson, Simon Dietz, and Eric Neumayer. Cheltenham: Edward Elgar.

Vogler, John. 2011. European Union Environmental Policy. In *Comparative Environmental Regionalism*, ed. Lorraine Elliott and Shaun Breslin. Oxon: Routledge.

Weale, Albert, Geoffrey Pridham, Michelle Cini, Dimitrios Konstadakopulos, Martin Porter, and Brandan Flynn. 2000. *Environmental Governance in Europe*. Oxford: Oxford University Press.

Williams, Marc. 2005. The Third World and Global Environmental Negotiations: Interests, Institutions and Ideas. *Global Environmental Politics* 5 (3): 48–69.

Young, Oran R. 1999. *Governance in World Affairs.* London: Cornell University Press.

Zürn, Michael. 1998. The Rise of International Environmental Politics: A Review of Current Research. *World Politics* 50 (4): 617–649.

CHAPTER 2

From Neo-liberalism
to Neo-extractivism

This chapter examines the context which made it possible for regional environmental cooperation in the Southern Cone to increase significantly from the 1990s onwards, as well as the constraints that explain why it has also remained marginal and has not become stronger. The chapter argues that this puzzle can only be understood with reference to the specific political and economic context of the region and in particular two simultaneous processes: the return to democracy which opened up political agendas and strengthened civil society on the one hand and the adoption of a development model based on export-driven growth and natural resource exploitation which leaves very little space for the consideration of environmental concerns on the other. The chapter examines the links between political and economic developments and socio-environmental concerns in two parts. The first part covers the 1990s and looks at the simultaneous impact of democratisation and neoliberal reforms. While the transition to democracy opened up political agendas and spaces and made the emergence of robust forms of regional environmental cooperation possible, by focusing on export-led growth and business interests the neoliberal reform agenda limited how far regional environmental cooperation could advance and thus accounts for its marginality. The second part then turns to the progressive governments that have come to power in a majority of South American countries since the turn of the century. These governments have introduced a number of new programmes and policies which have been successful in addressing urgent social concerns and reducing poverty. However, these achievements have relied to a large extent on income generated from the

© The Author(s) 2017
K.M. Siegel, *Regional Environmental Cooperation in South America*,
International Political Economy Series, DOI 10.1057/978-1-137-55874-9_2

exploitation and export of natural resources, and they have been facilitated by the commodity boom. As a result, natural resource exploitation has continued to gain economic and political significance. Resource exploitation has thus become increasingly intensive and extensive, and this has often had significant socio-environmental impacts. Yet, due to its importance in fulfilling electoral promises while keeping domestic and international elites involved in the export sector satisfied, governments have promoted and defended intensive resource exploitation and given little attention to the socio-environmental consequences. The neo-extractivist development strategy adopted by progressive governments has therefore deepened the old tension between development and sustainability, and this made it difficult for regional environmental cooperation to become stronger. However, governments have also continued to participate in global environmental processes, and there is a general recognition that environmental concerns need to be taken into consideration somehow even if there is no consensus on how exactly this should be done. In this context, robust forms of regional environmental cooperation have continued to develop, but remained in the margins of political agendas.

The 1990s: Opening up Space for Regional Environmental Cooperation While Keeping It in the Margins

The return to democracy during the 1980s changed the preconditions in such a way that increased and robust regional environmental cooperation became possible. Yet, at the same time during the 1990s, the central focus of policy-makers and international financial institutions was on export-driven economic growth with the assumption that this would also solve other concerns. Economic growth was to be achieved through a series of neoliberal reforms, notably opening up the markets and bringing in foreign investment while reducing the role and capacity of states. As a consequence of this reform agenda, the nascent forms of regional environmental cooperation remained marginal for several reasons. The shrinking of state budgets and responsibilities meant that although environmental institutions developed following democratisation, these were weak from the beginning. This is an important factor accounting for the high dependence on external funding. Furthermore, the priority given to export-led growth and natural resource exploitation made governments reluctant to develop stronger environmental agencies or adopt regulations or regional agreements which could present limitations to this growth strategy. This helps to understand why governments did not create regimes dedicated specifically to regional environmental concerns and why many agreements remain vague or non-binding. Finally, state–civil society relationships improved significantly with the end of the military dictatorships, but again this was shaped by the neoliberal reform agenda. Professional NGOs were thus able to gain prominent roles by providing services that the state no longer offered. Nevertheless, the extent to which civil society organisations could influence policy decisions in particular in relation to economic issues remained very limited. This means that although there were improvements in the relationship between states and civil society and this also benefitted regional environmental cooperation, the nature of the transition process restricted how far these could advance and explains why environmental cooperation remained marginal.

The return to democracy of the Southern Cone countries in the 1980s had at least four important implications which paved the way for the emergence of robust forms of regional environmental cooperation. As set out in the previous chapter, robustness relates to the ability of cooperation to continue even when the external circumstances change. The book argues that regional environmental cooperation in the Southern Cone can be deemed robust if regular activities of cooperation, what I call "cooperation in practice", become linked to agreements between governments, that is formal cooperation. First, democratisation opened up the political space for different civil society groups which under the dictatorships had been excluded from the policy-making process and persecuted if they were seen to be opposing the military regimes. The return to democracy meant that civil society organisations and protest movements could develop without having to fear government authorities, and this led to an expansion and strengthening of civil society organisations throughout the region (Grugel 2009, 32; Keck and Sikkink 1998, 130; Peters 2011). This included environmental groups which also increased and gained more possibilities for influencing political processes (Hochstetler 2012; Mumme and Korzetz 1997). Moreover, gradually, information regarding environmental concerns became available, allowing environmental groups to present a stronger case (Barton 1999, 199). Environmental groups were further strengthened by the Rio Summit in 1992 where Latin American NGOs had a strong presence. The summit also helped to promote the development of regional environmental NGO networks (Espach 2006, 64; Friedman et al. 2001; Hochstetler 2003, 26). Exchanges of information between different civil society organisations in the region and coordinated actions in relation to putting pressure on governments as well as carrying out conservation measures are all crucial elements of regional environmental cooperation in practice. This means the increased possibilities for civil society activity which democratisation brought and the changes in the relationship between the state and civil society also encouraged regional environmental cooperation.

Second, democratisation also opened up opportunities for restructuring existing institutions and including new issues on the political agenda, such as environmental concerns. Brazil was exceptional as it had already developed domestic environmental institutions under military rule. However, Argentina's military regime in fact abolished the environmental agency that had been created earlier, and neither Paraguay nor Uruguay created any domestic environmental institutions in that time period. This changed with the return to democracy, and during the 1990s, the Southern Cone countries significantly developed and strengthened their domestic environmental institutions and legislation (Hochstetler 2003, 2005, 353–356, 2012, 213–222). For example, in both Argentina and Paraguay, environmental concerns were included in constitutional reforms. Argentina's constitutional reform in 1994 set out that it is the responsibility of the national state to establish minimum standards for environmental protection whereas it is the responsibility of the provinces to pass more detailed corresponding legislation according to the characteristics of each province in order to implement the minimum standards (Bueno 2010, 124–126; Devia 1998; Di Paola and Rivera 2012, 16–18). In Paraguay, an environmental programme was set out in the new constitution of 1992 (Hochstetler 2003, 10; Díaz Labrano 1998), and a number of environmental laws relating to topics such as environmental impact assessments, eco-crimes, forestry or wildlife were passed in the 1990s. In addition, the start of the new millennium saw the creation of a new Environment Secretariat charged with the development of policies as well as the implementation of plans and programmes (Facetti 2002, 49–80). Uruguay developed basic legislation on environmental issues during the 1990s using loans from the Inter-American Development Bank (IDB), while Brazil introduced new legislation on environmental crimes in 1998 which raised the penalties for pollution considerably (Hochstetler 2003, 8–9).

Third, the new democratic governments were keen to gain recognition by the international community, and participating in international environmental politics and accepting international environmental norms was one way of achieving this. The wish to foster a certain state identity and become part of a group of states are powerful incentives to comply with dominant international norms (Finnemore and Sikkink 1998, 902–906). Moreover, international legitimation also increases domestic legitimacy, and both are particularly important in periods of domestic turmoil and uncertainty. States that are insecure with regard to their international recognition can thus be expected to have a high motivation to adhere to international norms. International pressure for environmental protection was particularly strong in the case of Brazil. As deforestation in the Amazon and the murder of the social and environmental activist Chico Mendes turned into a focus of concern in the global North during the 1980s, Brazilian politicians realised that in order to strengthen the country's international profile, they would have to demonstrate a commitment to international environmental norms. When Fernando Collor de Mello, Brazil's first democratically elected president, visited the capitals of countries in the industrialised North in early 1990 prior to his inauguration, he was thus confronted repeatedly with the question of how he would address "the Amazon problem" (Edwards and Roberts 2015, 44). It is not surprising then that in an effort to strengthen their position domestically and internationally, the newly democratic Southern Cone governments opened up to international environmental norms. Brazil sought to demonstrate its commitment to environmental concerns by hosting the 1992 UN conference in Rio de Janeiro, while Argentina hosted the UN climate change negotiations in 1998 and 2004 (Edwards and Roberts 2015, 44, 47; Hochstetler 2002, 40–41, 2005, 356). Both of these aspects, the development of domestic environmental institutions and the commitment to international environmental norms, are important in terms of promoting formal regional environmental cooperation, namely written and public agreements between governments in the region on shared environmental concerns.

Moreover, governments also became more open towards the involvement of international donors for environmental purposes, and since the 1990s, important environmental projects funded by international assistance have been implemented in all Southern Cone countries (Hochstetler 2005, 355–356). Projects with external donors are another important element which strengthens cooperation in practice, and in both case studies, this has become linked to formal cooperation. In the La Plata basin, from the 1990s onwards six large environmental projects have been carried out and all but one involved at least two countries. During the same time period, the riparian countries signed several agreements which included environmental concerns. In the case of the CMS, funding has come from many different sources and has been mostly channelled through environmental NGOs in the region. These have cooperated closely with governments and encouraged the signing of four memoranda of understanding between Southern Cone and some neighbouring countries under the umbrella of the CMS.

Finally, the return to democracy also encouraged more regional cooperation in general. The most obvious example of this was the creation of the regional organisation Mercosur in 1991, but the more cooperative climate also extended to other issues, such as joint projects in the La Plata basin (Elhance 1999, 35–36; Kaltenthaler and Mora 2002; Tussie and Vásquez 2000). This also had repercussions for environmental cooperation. The number of treaties with environmental components signed between South American countries thus increased substantially in the 1990s (see Fig. 2.1). Although individual initiatives of course existed before, overall the early 1990s thus marked the start of increasingly robust regional environmental cooperation.

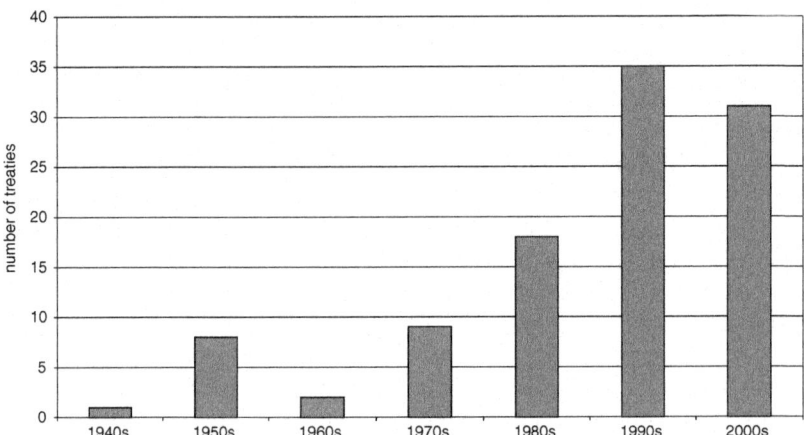

Fig. 2.1 Number of treaties with environmental components signed between South American countries per decade. *Source* Developed by author based on data provided in the ECOLEX database (data extracted in November 2010)

However, the 1990s were shaped not only by the process of democratisation, but also by neoliberal reforms. At the global level, the 1992 Earth Summit institutionalised norms of liberal environmentalism which linked environmental protection closely to economic growth, free trade and market-based mechanisms as the preferred instruments to address environmental concerns (Bernstein 2000). Throughout South America, structural adjustment packages were promoted by governing elites under strong pressure from international financial institutions, notably the World Bank and the International Monetary Fund (IMF) as well as the US government (Gwynne and Kay 2000). The central idea was that market opening while reducing state spending would lead to economic growth and at the same time strengthen democracy (Grugel 2009, 33). Democratisation thus took place in the context of neoliberal reforms, and this significantly shaped the nature of the transition process. It also meant that although following democratisation robust forms of regional environmental cooperation developed in several areas, these could not gain more than a marginal position in regional cooperation. As set out in Chap. 1, marginality of regional environmental cooperation is characterised by an absence of new regimes created specifically for regional environmental concerns, a high dependence on external funding and the vague or non-binding nature of agreements. This section examines the different ways in which neoliberal reforms and the nature of the transition process contributed to the marginalisation of the nascent forms of regional environmental cooperation in the Southern Cone.

First, a central element of the neoliberal reform agenda was the reduction of state capacity and responsibilities in an effort to cut down on government spending and reduce corruption. This went hand in hand with a strengthening of the private sector and in particular large-scale transnational corporations. This was closely related to a second important element of the neoliberal reform agenda which was the focus on export-led growth and natural resource exploitation. Governments in the Southern Cone and all over South America thus carried out a series of measures aimed at opening up the markets, cutting external tariffs and bringing foreign investors into sectors of the economy that had been closed up till then, including public utilities and natural resources (Grugel 2009, 32–33; Grugel and Riggirozzi 2012, 4; Gwynne and Kay 2000). This was very evident in the mining, oil and gas sectors where a series of reforms, including privatisation, lower taxes and more labour flexibility, were implemented with the aim of attracting investors and sometimes complemented with agreements guaranteeing extremely favourable conditions (Hogenboom 2012, 135–140). Neoliberal policies also extended to other types of natural resources and in many countries included the privatisation of water management, both in relation to rivers and hydropower installations and in relation to sanitation and the provision of drinking water (Wickstrom 2008). Across the global South, neoliberal reforms also contributed to the depoliticisation and technocratic governance of resources as well as increasing the influence of international actors (Grugel and Nem Singh 2013, 67). As discussed in more detail below, economic reforms also affected agriculture and forestry, two sectors which are specific to the large plains and rivers of the Southern Cone, but less prominent in other parts of South America. Overall, using their comparative advantage in this area, South American countries thus relied increasingly on exports of commodities for economic growth (Green 1999; Murray 1999) although this trend was less prominent in Brazil which had a more developed industrial base and a lower reliance on exports for growth (Burges 2009).

A context where natural resource exploitation, export-led growth and a strengthening of business interests were central elements in the development model adopted was not favourable to the development of stronger forms of regional environmental cooperation. While neoliberal policies prioritised economic growth and business interests, environmental regulations were seen as obstacles which might limit growth or lead to opposition by business. In a development model oriented to such an extent towards economic growth, there was little political will to create strong environmental agencies or develop more stringent regulations (Barton 1999, 195–196). This also demonstrates how international factors shaped the region's environmental politics in contradictory ways. As Hochstetler and Keck point out in relation to Brazil, while foreign environmentalists and politicians demanded that the authorities protect the Amazon rainforest, international financial institutions simultaneously insisted on reducing state resources and responsibilities (Hochstetler and Keck 2007, 8). This meant that, although following democratisation, governments in the Southern Cone and all over South America, created new environmental institutions, these never became very strong and often did not have the capacity to carry out their mandates. Moreover, they remained considerably less powerful than other state agencies and ministries in terms of their budgets and resources (Barton 1999, 195; Gwynne and Silva 1999, 159–160; Mumme and Korzetz 1997, 53–54). Furthermore, although countries in the region strengthened their environmental legislation, serious shortcomings remained. This included the adoption of legislation primarily for symbolic reasons as well as a lack of implementation and high barriers for citizens and environmental groups who wished to refer to environmental legislation to build a court case (Mumme and Korzetz 1997, 51–52). All of this seems to suggest that the commitment to the newly created environmental agencies and regulations was only half-hearted and perhaps driven less by a concern to address environmental problems effectively than to improve the reputation of governments at the domestic and international level. Overall, the opening that democratisation brought was therefore important in making increased and more robust forms of regional environmental cooperation possible, but the particular conditions in which the transition took place were unfavourable for the creation of stronger forms of regional environmental cooperation and in some cases this had direct impacts on cooperation. The initial report for a shared project on the Rio de la Plata between Argentina and Uruguay, for example, noted as one of the problems the reduced capacity of state agencies for monitoring and controlling environmental problems in the river as a result of the state reforms of the 1990s (FREPLATA 2005, 206).

Third, the nature of the transition process also shaped state–civil society relations in a particular way, and this is one of the elements which influences the level of robustness of regional environmental cooperation. While governmental agreement is necessary for formal agreements, civil society organisations are important drivers for cooperation in practice. With the end of the military dictatorships and repression, state–civil society relations changed considerably. Most notably, the antagonism between civil society and state authorities which had characterised the dictatorships started to break up and was replaced by new, more complex relationships. In particular, professional NGOs gained prominent positions in neoliberal democracies and increasingly worked together with governments to provide services that the state no longer offered (Taylor 1999; Tedesco 1999). To some extent, cooperation between governments and NGOs thus improved, and this is important to link formal agreements with regular activities and achieve robust forms of regional environmental cooperation. However, the reliance on external funding and sources of support for regional environmental cooperation is also one of the key reasons for its marginality.

Moreover, there were again limitations in how far state–civil society relations could improve in the context of neoliberal reforms. In fact, promoting NGOs as service providers has meant that organisations focussed more on solving practical problems than political activity or ideological questions (Taylor 1999, 142). In addition, although the assumption was that liberalised markets would strengthen democracy, neoliberal reforms were in fact carried out in a top-down manner with very little popular consultation. Even though democratisation strengthened civil society, this rarely led to any real influence over the political agenda, in particular in relation to economic questions (Grugel 2001, 165–190; Grugel 2009, 32–33). Instead, economic reforms were carried out on the basis of a consensus between domestic and international economic and political elites. Domestic and transnational business groups who were involved in the export sector or the privatisation processes were important sources of investment and legitimacy for governments and consequently had good access to decision-makers while other groups and their points of view were sidelined (Grugel 2009, 36; Phillips 1999, 86; Phillips and Buxton 1999, 2–3). Many South American citizens, on the other hand, saw the privatisation of natural resources as unfair as they believed that the mineral wealth of their countries should benefit the people rather than foreign investors (Hogenboom 2012, 137–139; Perreault 2008). Moreover, it was the elites who mostly reaped the benefits of economic growth, while the poor suffered from increasing unemployment and insecurity as well as cutbacks in state welfare programmes. Latin America as a whole thus averaged poverty rates of about 40% during the 1990s, and in Argentina, unemployment rates reached almost 20% by 2002 (Green 1999, 22; Grugel and Riggirozzi 2012, 4–5). While there are important differences between countries, the neoliberal reforms of the 1980s and 1990s have deepened inequality and economic and political exclusion in many cases and reinforced distrust of the state and disappointment with the way democracy functioned (Bull 2013, 80–81; Grugel 2001; Gwynne and Kay 2000).

On the whole, this means that state–civil society relations did open up, but in a very selective way with deep divisions remaining. Civil society organisations and in particular professional NGOs were able and encouraged to take over some of the functions of the state, but decision-making remained dominated by economic and political elites in particular in relation to economic questions, such as natural resource governance. This meant the possibilities of civil society to promote stronger environmental regulations or more specific regional agreements remained limited. Overall, democratisation was thus crucial in setting the preconditions for the development of robust forms of regional environmental cooperation, but the nature of the transition process restricted how far these could advance and explain why they remained marginal.

Reinforcing the Marginality of Socio-environmental Concerns: The Neo-extractivist Development Model

Social unrest and protests against the neoliberal reform agenda had simmered throughout the 1990s and erupted spectacularly in Argentina in 2001 in the midst of hyperinflation and an economic crisis that also impacted on neighbouring countries. Protestors demanded that their interests also be represented by political leaders and governments and asked for a new democratic pact and a different model of citizenship (Grugel 2009, 37–42). Middle-class and low-income protestors united under the slogan "Que se vayan todos" (Out with all of them) and toppled several governments over the course of only a few weeks (Grugel and Riggirozzi 2007; Prevost 2012). Meanwhile in neighbouring Bolivia, the governance of natural resources became the focus of popular opposition. Here, protests against the privatisation of water in the city of Cochabamba in 2000, the so-called water wars, became so strong that they succeeded in reversing the policy. Three years later, widespread protests shook the whole country and brought down the president at the time, Gonzalo Sánchez de Lozada. The cause this time was the privatisation of gas which the government defended heavy-handedly and which gave extremely favourable conditions to foreign investors, but very few benefits for the Bolivian state and population. These events paved the way for the election of Evo Morales, the country's first indigenous president, in 2005 (Crabtree 2009, 95–96; Perreault 2008; Wickstrom 2008). In the course of the 2000s, dissatisfaction with the policies of previous governments brought Leftist governments to power in a majority of South American countries and in all the Southern Cone countries. Consequently, in 2009 about three-quarters of South America in terms of territory and 80% of its population was governed by various Leftist governments (Gudynas 2009, 189), but with political changes in Argentina and Brazil in 2015 and 2016, the left tide has been weakened significantly.

In Argentina, Néstor Kirchner brought the Peronist Party back to its original centre-left position and won the presidential elections of 2003. Following two election victories of his wife Cristina Fernández de Kirchner of the same party in 2007 and 2011, the Kirchner era came to an end in 2015 with the election of centre-right leader Mauricio Macri as the president. In neighbouring Paraguay, the election of Fernando Lugo, a former bishop and defender of liberation theology[1] in 2008 ended over 60 years of conservative rule by the Colorado Party. However, Lugo's presidency came to a premature end with a contested impeachment process in 2012 in which disputes relating to land rights and natural resource governance played a key role, as discussed in the following chapter. In the region's largest country, Brazil, another controversial impeachment process in 2016 ended 14 years of centre-left government. Luiz Inácio da Silva (Lula), the leader of the Worker's Party and a former industrial worker, won the presidency in 2002 and was re-elected for a second term in 2006. Not being able to run for president a third time, Lula then supported Dilma Rousseff who won the presidential elections in 2010 and, by a small margin, again in 2014. However, in the midst of a political and economic crisis including a sprawling corruption scandal involving politicians of all political parties at the same time as a fall in commodity prices devaluing some of the country's key exports, the Rousseff government came under severe criticism in the first half of 2016. Following a highly controversial process that has been widely criticised for its political motivations, Rousseff was impeached in August 2016 and a new government was formed under Michel Temer. Since then protests have continued and corruption investigations are ongoing involving several of the politicians who replaced Rousseff's government. Finally, Uruguay remains under a progressive government as the Frente Amplio coalition bringing together several Leftist parties has been in power since 2004 under the presidents Tabaré Vásquez and José Mujica.

There are many differences and nuances in the positions and policies of the left tide governments that shaped domestic and regional politics during the first 15 years of the new millennium, but crucial shared elements include their strong criticism of the neoliberal reform agenda and their commitment to addressing social concerns and reducing poverty (Cannon and Kirby 2012; Grugel and Riggirozzi 2012; Panizza 2009, 168–196). What is more, by using the income generated from the exploitation and export of natural resources and the favourable economic context offered by the commodity boom in order to achieve social objectives, they adopted very similar neo-extractivist development strategies even if there are of course variations between countries and sectors. As a result, natural resource exploitation has become highly significant politically and economically for progressive governments and intensive resource exploitation also looks set to continue under conservative governments. This has deepened the old tension between development and sustainability and hindered a strengthening of socio-environmental concerns on domestic and regional political agendas. Nevertheless, governments have also continued to participate in global environmental processes, and there is a general recognition that environmental concerns need to be taken into consideration somehow even if there is no consensus on how exactly this should be done. Over the last 15 years, robust forms of regional environmental cooperation have thus continued to develop, but remained in the margins of political agendas.

The high global prices for commodities during the 2000s sparked by increasing demand from Asia and China in particular provided a highly favourable economic context for resource-exporting South American countries. It allowed governments to adopt more assertive foreign policies and China's interest in the region provided more room for policy autonomy (Edwards and Roberts 2015, 34; Fernández Jilberto and Hogenboom 2010). As a result, natural resource exploitation has not only continued, but has become increasingly intensive and extensive. Increases in natural resource exploitation are evident in the volumes extracted, the share of commodities relative to other exports and the extent of areas used for resource exploitation. While there are differences between countries and sectors, there is a clear trend towards "reprimarisation" and economies of extraction (Burchardt and Dietz 2014, 427; Svampa 2012, 17). What is more, in the absence of other development strategies, for example changes to the rather regressive taxation systems in place across South America, neo-extractivism shows signs of having become a consolidated development project rather than just a temporary economic strategy (Burchardt and Dietz 2014).

Similar to South America as a whole, all the Southern Cone governments have promoted increased natural resource exploitation in order to achieve economic growth through commodity exports. Although it has the most diversified economy and export sector in the region, this trend is also valid for Brazil. As a consequence of the growing importance of China as a trading partner, the proportion of primary products in Brazil's total exports thus increased from 22.8 to 43.4% during the first decade of the new millennium (Hochstetler 2013, 40, 42). Neo-extractivism also includes agricultural production, notably soybean, because it shares many of the key characteristics of the traditional extractive industries. These are large-scale extraction processes to export high volumes of resources with very little or no processing (Gudynas 2010a, 40). In the Southern Cone, natural resource exploitation has increased in three key sectors.

First, agricultural production, which has always been one of the main sectors of the economies in the Southern Cone, has changed significantly since the neoliberal reforms of the 1980s and 1990s. Following market liberalisation, large agribusinesses increasingly took over agricultural production in sectors where profits could be expected (Gudynas 2008, 514; Newell 2008, 347; Robinson 2008, 88). In large parts of Argentina, Bolivia, Brazil, Paraguay and Uruguay, this led to an explosion of soybean plantations since the 1990s. Soy exports have become highly profitable because of high global demand for biodiesel as well as animal feed with rising demands in particular from China. Soybean has thus become the most important agricultural export of the Southern Cone countries and the region has become a leading producer and exporter of the crop worldwide (Gudynas 2008, 513; Lapitz et al. 2004; Newell 2008, 347–348; Robinson 2008, 84). Progressive governments have actively supported the expansion of agribusiness by introducing various state institutions, policies and incentives designed to promote the sector (Fulquet 2015; Rivera-Quiñones 2014; Vergara-Camus 2015).

Forest monocultures and the production of pulp for the paper industry which has expanded in parts of Argentina, Brazil, Chile and Uruguay have followed a similar path. This development was initiated by neoliberal policies, but also continues under progressive governments, as the cases of Chile and Uruguay demonstrate. In Chile, the neoliberal policies of the Pinochet dictatorship started to make the forestry industry into one of the main pillars of the economy. This continued also under democratic governments, and since the 1990s, the sector has expanded exponentially supported by the IMF and the World Bank and promoted with policies of privatisation as well as direct and indirect subsidies, such as tax exemptions (Cuenca 2005; Ortiz et al. 2005, 44–46). Uruguay followed a similar path about a decade later and has since become a new key location for the pulp industry. Under the conservative government of President Sanguinetti, the first elected president after the military dictatorship, the country approved a law on forestry in 1987 declaring the forestry sector a national interest and establishing extremely favourable conditions for foreign investors including state subsidies and tax exemptions. These were reinforced when a bilateral investment agreement was signed between Uruguay and Finland in 2002. Overall, since the 1980s, both monocultures of trees for the industry and the construction of pulp mills have increased significantly in the region (Chidiak 2012; Gutiérrez and Panario 2014).

Finally, reflecting developments in the Andean countries, the mining sector too has continued to expand under progressive governments in the Southern Cone. This manifests itself in increased production in countries such as Brazil and Argentina as well as moving into new areas. These include frontier areas, such as the large-scale Pascua Lama project between Argentina and Chile and countries where mining is being established as a new sector, such as Uruguay (Gudynas 2009, 191; Saguier 2012a, 127). Again, these trends started during the 1990s; Argentina and Chile, for example, signed a bilateral mining treaty in 1997 (Saguier 2012a, 129, b) and continued during the 2000s. In Argentina, for example, investments in mining went up by 740% between 2003 and 2009 (Saguier 2012a, 127).

Intensive resource exploitation has become highly significant politically because it has allowed progressive governments to respond to social demands as well as elite interests. Neo-extractivism combining old practices of natural resource exploitation with new social policies and a stronger role of the state has thus become the dominant development strategy over the past decade (Burchardt and Dietz 2014; Gudynas 2009). Domestically, revenues from the commodity sector have been a central element that made social programmes possible and allowed progressive governments to address some of their most urgent priorities. After the 2001 crisis in Argentina, export taxes on agricultural commodities and hydrocarbons became crucial for social emergency programmes (Grugel and Riggirozzi 2007, 96; Riggirozzi 2009, 104) and soy exports were a key element in Argentina's impressive economic recovery after the crisis at the start of the millennium (Newell 2009, 51; Robinson 2008, 86). Since then, the revenues generated by agricultural exports have continued to play a central role in providing the Kirchner governments with the fiscal space needed for social programmes (Rivera-Quiñones 2014). Moreover, the high commodity prices and increased revenues helped Brazil and Argentina to pay back their debts to the IMF before the deadlines, thus making them independent of the policy prescriptions of international financial institutions (Hogenboom 2012, 149; Hogenboom and Fernández Jilberto 2009, 94). In Argentina, the reduced debt burden in combination with the commodity boom helped to advance on important social issues and increase spending on health, education and housing (Prevost 2012, 27–28). Similarly, in Uruguay and Brazil, the commodity boom and the strengthened economy provided the necessary favourable context to expand social programmes (Gudynas 2009, 208; Zibechi 2010, 107–108). One of the most well-known cases is the Brazilian Bolsa Família programme that extended social spending to over 11 million families. It offered financial support linked to the condition that children attend school (Branford 2009, 161; Burges 2009, 207). Social programmes like these have been crucial for progressive governments to keep their election promises and provided the basis of their political legitimacy and popularity (Gudynas 2009, 213). Export-led growth and natural resource exploitation have thus become the basis for economic and social development, and this has resulted in some significant achievements. Between 1990 and 2010, poverty rates in Latin America decreased significantly from half of the population living in poverty to a third (Burchardt and Dietz 2014, 473–474; Edwards and Roberts 2015, 33). This success has provided an important pillar of the legitimacy and popularity of progressive governments (Gudynas 2010b, c; Hogenboom 2012; Hogenboom and Fernández Jilberto 2009). With the exception of Paraguay, progressive governments were re-elected in all the Southern Cone countries often with very high approval rates. This contrasted starkly with the political instability in the 1990s and early 2000s, particularly in the Andean region, when presidents regularly did not finish their terms in office (Bull 2013, 81–83) although the impeachment processes in Brazil and Paraguay are worrying indications of continuing deep divisions that weaken stability and democracy.

The promotion of intensive natural resource exploitation has allowed progressive governments to reconcile some of these divisions at least temporarily and respond to the demands of domestic and international economic elites and in particular the business groups linked to the export sector. As Grugel argues, Latin America has seen much less of a transformation of elite mentalities than Europe. Instead, elites have been much more reluctant to accept the notion that the state should assume responsibility in relation to the welfare of citizens and that citizens have social and economic as well as political rights (Grugel 2007, 247–248). This means elite opposition can pose serious obstacles for progressive policies (Riggirozzi and Grugel 2009, 222–223), so that it is crucial for Leftist governments to secure at least a minimum of support. At least during the first decade of the 2000s when high commodity prices provided a favourable economic context, neo-extractivism as a development strategy offered ways of making some progress on social concerns while responding to elite interests. This is particularly clear, for example, in the case of Bolivia where the Morales government initially faced strong challenges from traditional landowning elites producing soya in the Media Luna region of Bolivia who regarded Morales as a threat to the export of agricultural commodities. Yet, despite some continuing tensions, over time, the two sides have reached a consensus as the Morales government did not challenge large landholdings and tacitly came to support genetically-modified (GM) soybean production. This also benefitted a new emerging elite of small farmers supportive of the government who also turned towards GM soya production due to the economic profits, the absence of an alternative market and the difficulties in maintaining non-GM agricultural production in an international political economy heavily favouring GM production. The support for GM soy production for export has thus allowed Morales to appease traditional landowning elites challenging his government as well as strengthen an emerging new elite of small farmers supporting his party (Høiby and Zenteno Hopp 2015).

The adoption of neo-extractivism as the main development strategy has further deepened the tension between achieving economic and social development and socio-environmental sustainability and kept environmental concerns in the margins of political agendas. Due to the socio-environmental impacts, projects of resource exploitation have been challenged in contestations and protests across South America, but as I have argued elsewhere (Siegel 2016), this has been met by governments with a combination of discourses justifying intensive resource exploitation and internal prioritisation of resource exploitation over other concerns in policy-making and implementation. As a result, environmental policies have often remained inconsistent and ineffective.

The approach of progressive governments towards resource governance has also shaped relations between the state, civil society and business in ways that were perhaps unexpected within parts of the Left and that further contributed to the marginalisation of socio-environmental concerns and those questioning the sustainability of the development model. Many of the South American Leftist presidents have their roots in social movements and have come to power with the support of social movements including indigenous organisations, peasant associations and other groups seeking environmental justice and a fairer and more sustainable use of natural resources (Bull 2013, 75; Bull and Aguilar-Støen 2015, 2; Prevost et al. 2012, 12–14). Yet, once in power relations between progressive governments and social movements have evolved in complex and sometimes unexpected ways. In several countries, divisions and new lines of conflict have opened up in particular in relation to socio-environmental concerns and natural resource governance. As Gudynas points out, before coming to power many on the Left frequently criticised the social and environmental impact of natural resource exploitation and economic development based on export-driven growth. Reasons for this criticism included the local impact of extractive industries and agribusiness, the poor working conditions or the marginalisation of peasants and small agricultural producers vis-à-vis the growing power of large and often foreign companies, as well as economic arguments, notably the dependency on exports and the vulnerability to world market prices of commodities over which the exporting countries have little influence (Gudynas 2009, 188–189, 2010a, 38–39, b, 66). The election of progressive governments thus sparked hopes that this would be the start of new and alternative approaches to development and environment with more emphasis on sustainability and environmental justice and moving away from resource extraction (Bull and Aguilar-Støen 2015, 2; Hogenboom 2015, 127; Silva 2016, 327–328). Symbolically environmental concerns did gain in importance in many countries (de Castro et al. 2016, 3), and in Bolivia and Ecuador, a different approach to nature, well-being and development invoking indigenous concepts was also codified in new constitutions (Lewis 2016, 176–180). Yet, in practice the adoption of neo-extractivism as a development strategy has meant that natural resource governance under progressive governments across the region has often been riddled with contradictions and reversals of the positions held prior to elections (Bull and Aguilar-Støen 2016, 139–141).

In Brazil, for example, Lula had previously criticised GM technology which transnational agricultural companies strongly pushed for, but this was adopted under his presidency (Gudynas 2010a, 38; Hochstetler and Keck 2007, 180). Moreover, Lula's presidency turned out disappointing for the Landless Workers Movement (Movimento Sem Terra or MST) who had supported him in the 2002 election. Protesting against the highly unequal land distribution in Brazil, the MST's main focus is agrarian reform, but in response to the neoliberal reform agenda, it has broadened out to include demands in opposition to national and global capitalist expansion. Although the MST has rejected direct participation in party politics, it created close ties to the Worker's Party and supported Lula, not least by promising not to engage in new land occupations during the election campaign in order to reassure middle-class voters (Bull 2013, 88–89). Following his election, Lula introduced a few changes, but overall his approach to the question of land reform was far more moderate than he had advocated previously (Branford 2009; Newell 2008; Vanden 2012). In Uruguay, the Left-wing Frente Amplio coalition sided with environmentalists while in opposition and criticised the construction of a large pulp mill on the Uruguay River (Berardo and Gerlak 2012, 107; Pakkasvirta 2010, 77), but once in government it did not fundamentally alter policies in relation to the pulp industry. In 2005, the new Left-wing government thus made minor changes to the neoliberal policies from the previous decade and stopped direct subsidies to the sector, but the tax exemptions remained in place. Moreover, the construction plans of new pulp mills in Uruguay have continued also under progressive governments (Ortiz et al. 2005).

This means that in many cases projects of resource exploitation have been contested by citizens or movements who had previously supported the very governments who then carried out the projects criticised. State–civil society relations under progressive governments have thus moved into a new phase (Kirby and Cannon 2012, 11), and a "shift of conflicts" (Hogenboom 2012, 151) has taken place. This is particularly pronounced in Ecuador where, on the one hand, a stronger state created under Correa's government has meant a strengthening of environmental institutions and this has also resulted in the loss of influence of transnational funders. However, on the other hand, this stronger state continues to rely on intensive exploitation of subsoil resources, and there is mounting evidence of suppression of civil society groups opposing projects of resource exploitation. Consequently, some activists have become increasingly concerned that democracy is being eroded (Lewis 2016, 162–194). Overall, across South America, in particular groups advocating alternative and more radical changes in the approach to development and environment found that their hopes for greater changes under progressive governments had been dashed (Silva 2016, 329–330).

At the same time, domestic and foreign companies seeking to increase large-scale intensive resource exploitation have been able to expand their position and continue to exert significant political influence often to the detriment of stronger environmental regulations and environmental justice demands. This is evident particularly in the case of the agribusiness sector which has been successful in building strong ties with policy-makers in the Southern Cone countries. In Argentina, for example, biotechnology corporations enjoy close links to the government and formal access to decision-making. Together with their enormous material resources, this has put transnational companies in a strong position to shape political agendas. Moreover, biotechnology corporations have used their influence and resources not only to shape policy-making, but also to access and sponsor mass media (Newell 2009). The position of transnational companies and Monsanto in particular has been further strengthened by the US embassy which, according to Wikileaks data, promoted their interests in Argentina (Zenteno Hopp et al. 2015, 75, 86). The agribusiness lobby is also strong in Brazil and has been able to exert significant influence on government policies and resist stronger environmental regulations (Branford 2009, 159; Hochstetler 2013, 41; Vanden 2012, 44). In 2012, for example, the pro-agribusiness group in the Brazilian Congress pushed for changes to the Forest Code, the legislation designed to protect forests. The changes have been heavily criticised by environmentalists and demonstrate the lack of consensus regarding the need for forest protection in relation to export agriculture (Edwards and Roberts 2015, 90). Overall, Brazil's environmental management of industrial production has thus been more successful than of the natural resources and agricultural sectors (Hochstetler 2013, 40; Hochstetler and Keck 2007). In Uruguay too, the export boom has led to the strengthening of multinational corporations resisting political changes, and the widespread use of biotechnology with all soybean grown in Uruguay being genetically modified (Thimmel 2010, 104; Zibechi 2010, 110).

On the whole, hopes that socio-environmental concerns would gain more weight under progressive governments did not materialise and in particular environmental justice concerns such as land rights and distribution of resources have been largely off the political agenda. Moreover, implementation of existing regulations often remains a major challenge with potentially very severe consequences. This was demonstrated very clearly by the social and environmental catastrophe that unfolded in Brazil at the end of 2015 when a mining dam collapsed and caused a toxic mudslide that killed several people, left hundreds homeless and thousands without access to clean drinking water, not to mention the impacts on land and coastal ecosystems as the mud-slide made its way to the sea.

In many ways, neo-extractivism has thus reinforced the marginality of environmental concerns on policy agendas as well as the sidelining of those promoting them in policy processes. This has been particularly evident in relation to environmental concerns connected to resource exploitation. However, this has to be seen in the context of the constraints imposed by a global capitalist economy in which South American states generally do not have a very strong position and which favours continuing resource exploitation in South America as discussed in the final chapter. Yet, this does not mean that progress made previously has been reversed. On the contrary, as Silva argues, despite significant obstacles, the institutional capacity, technical knowledge and the national and international networks of professional staff have increased in Latin American countries without a doubt over the past 30 years (Silva 2016, 328–329). Global environmental processes also continue to play a role. Brazil thus hosted a follow-up of the Rio Summit in 2012 even though it was not proactive in terms of pushing the summit forward (Edwards and Roberts 2015, 99). Despite the limitations and contradictions, South American progressive governments have therefore maintained a commitment to environmental concerns in line with the globally dominant "green economy" paradigm (de Castro et al. 2016, 9) which emphasises market-based mechanisms and technical solutions to deal with environmental concerns, but avoids larger questions of social and environmental justice or wider participation in decision-making procedures. There is also some evidence of the emergence of new, environmental technocratic elites operating within the green economy framework which are at times able to exert some influence despite facing important constraints, but there is also a danger that alternative approaches and environmental justice concerns are sidelined in the process (Aguilar-Støen and Hirsch 2015; Bull and Aguilar-Støen 2016; Parker G. 2015; Toni et al. 2015). It is in this context that the Southern Cone countries have signed new agreements and declarations on shared environmental concerns, for example in relation to the Guaraní aquifer and various groups of migratory species as discussed in Chaps. 4 and 5. Moreover, domestic civil society groups, networks of researchers and some government officials as well as transnational funders have continued to work towards regional environmental cooperation. This means robust forms of regional environmental cooperation have continued to develop, but these are still marked by significant limitations as regional environmental cooperation remains highly dependent on external support and agreements that have been signed by progressive governments also lack specific guidelines and commitments. The continuing marginality is to a large extent an outcome of the development path taken and the continuing tension between achieving developmental objectives and environmental ones. Over the time period examined, intensive resource exploitation with significant socio-environmental consequences has become a cornerstone of neoliberal and progressive governments. The shift of priorities towards social concerns under progressive governments has led to important achievements, but the reliance on increasingly intensive and extensive resource exploitation meant that socio-environmental concerns have remained clearly subordinate to objectives of economic and social development. The next chapter outlines the transboundary dimensions of neo-extractivism and how this has also shaped regional integration and relations between states at the regional level. Because resource governance is also central to regional integration processes, it would be difficult for governments to promote regional environmental cooperation, but leave out the central issue of the socio-environmental consequences of resource exploitation. As a consequence, regional environmental cooperation has come to develop largely separate from government-led regional integration processes and regional organisations. This is particularly striking in the case of the regional organisation Mercosur which initially seemed like a promising framework for regional environmental cooperation, but whose relevance continuously declined over time.

NOTE

1. Liberation theology is a progressive movement of the Roman Catholic Church that started in Latin America in the 1960s and promotes profound reforms to address the causes of political, economic and social injustice. For Lugo's role in the Catholic Church, see O'Shaughnessy and Ruiz Díaz (2009).

REFERENCES

Aguilar-Støen, Mariel, and Cecilie Hirsch. 2015. REDD+ and Forest Governance in Latin America: The Role of Science-Policy Networks. In *Environmental Politics in Latin America—Elite Dynamics, the Left Tide and Sustainable Development*, ed. Benedicte Bull and Mariel Aguilar-Støen. Oxon: Routledge.
Barton, Jonathan R. 1999. The Environmental Agenda: Accountability for Sustainability. In *Developments in Latin American Political Economy—States, Markets and Actors*, ed. Julia Buxton and Nicola Phillips. Manchester: Manchester University Press.
Berardo, Ramiro, and Andrea K. Gerlak. 2012. Conflict and Cooperation Along International Rivers: Crafting a Model of Institutional Effectiveness. *Global Environmental Politics* 12 (1): 101–120.
Bernstein, Steven. 2000. Ideas, Social Structure and the Compromise of Liberal Environmentalism. *European Journal of International Relations* 6 (4): 464–512.
Branford, Sue. 2009. Brazil: Has the Dream Ended? In *Reclaiming Latin America: Experiments in Radical Social Democracy*, ed. Geraldine Lievesley and Steve Ludlam. London: Zed Books.

Bueno, María del Pilar. 2010. *De Estocolmo a La Haya – La Desarticulación de Las Políticas Ambientales En La Argentina*. Rosario: UNR Editora – Editorial de la Universidad Nacional de Rosario.

Bull, Benedicte. 2013. Social Movements and the 'Pink Tide' Governments in Latin America: Transformation, Inclusion and Rejection. In *Democratization in the Global South*, ed. Kristian Stokke and Olle Törnquist. Basingstoke and New York: Palgrave Macmillan.

Bull, Benedicte, and Mariel Aguilar-Støen. 2015. Environmental Governance and Sustainable Development in Latin America. In *Environmental Politics in Latin America—Elite Dynamics, the Left Tide and Sustainable Development*, ed. Benedicte Bull and Mariel Aguilar-Støen. Oxon: Routledge.

Bull, Benedicte, and Mariel Aguilar-Støen. 2016. Changing Elites, Institutions and Environmental Governance. In *Environmental Governance in Latin America*, ed. Fabio de Castro, Barbara Hogenboom, and Michiel Baud. Basingstoke: Palgrave Macmillan.

Burchardt, Hans-Jürgen, and Kristina Dietz. 2014. (Neo-)extractivism—A New Challenge for Development Theory from Latin America. *Third World Quarterly* 35 (3): 468–486.

Burges, Sean W. 2009. Brazil: Toward a (Neo)Liberal Democracy? In *Governance after Neoliberalism in Latin America*, ed. Jean Grugel and Pía Riggirozzi. New York: Palgrave Macmillan.

Cannon, Barry, and Peadar Kirby. 2012. Civil Society—State Relations in Left-Led Latin America: Deepening Democratization? In *Civil Society and the State in Left-Led Latin America—Challenges and Limitations to Democratization*, ed. Barry Cannon and Peadar Kirby. London: Zed Books.

Chidiak, Martina. 2012. Investment Rules and Sustainable Development: Preliminary Lessons from the Uruguayan Pulp Mills Case. In *Rethinking Foreign Investment for Sustainable Development Lessons from Latin America*, ed. Daniel Chudnovsky and Kevin Gallagher. Cambridge: Cambridge University Press.

Crabtree, John. 2009. Bolivia: Playing by New Rules. In *Reclaiming Latin America: Experiments in Radical Social Democracy*, ed. Geraldine Lievesley and Steve Ludlam. London: Zed Books.

Cuenca, Lucio. 2005. Celulosa Arauco En Valdivia. El Desastre Ambiental En El Río Cruces, Resultado Del Modelo Forestal Chileno. In *Entre El Desierto Verde Y El País Productivo: El Modelo Forestal En Uruguay Y El Cono Sur*, ed. María Selva Ortiz, Javier Taks, Beatriz Schmid, and Stefan Thimmel. Montevideo: Casa Bertolt Brecht and REDES-Amigos de la Tierra. http://www.redes.org.uy/wp-content/uploads/2008/10/entre-el-desierto-verde-y-el-pais-productivo.pdf.

De Castro, Fabio, Barbara Hogenboom, and Michiel Baud. 2016. Introduction: Environment and Society in Contemporary Latin America. In *Environmental Governance in Latin America*, ed. Fabio de Castro, Barbara Hogenboom, and Michiel Baud. Basingstoke: Palgrave Macmillan.

Devia, Leila (ed.) 1998. *Mercosur Y Medio Ambiente*, 2a ed. Buenos Aires: Ediciones Ciudad Argentina.

Di Paola, María Marta, and Inés Rivera. 2012. Informe Nacional Sobre El Estado Y Calidad de Las Políticas Públicas Sobre Cambio Climático Y Desarrollo En Argentina. Buenos Aires. http://www.intercambioclimatico.com/articulos/.

Díaz Labrano, Roberto Ruiz. 1998. La Defensa Y Preservación Del Medio Ambiente En El Ordenamiento Jurídico Del Paraguay. In *Mercosur Y Medio Ambiente*, 2a ed., ed. Leila Devia. Buenos Aires: Ediciones Ciudad Argentina.

Edwards, Guy, and J. Timmons Roberts. 2015. *A Fragmented Continent—Latin America and the Global Politics of Climate Change*. Cambridge, MA: MIT Press.

Elhance, Arun P. 1999. *Hydropolitics in the Third World—Conflict and Cooperation in International River Basins*. Washington, D.C.: United States Institute of Peace Press.

Espach, Ralph. 2006. When Is Sustainable Forestry Sustainable? The Forest Stewardship Council in Argentina and Brazil. *Global Environmental Politics* 6 (2): 55–84.

Facetti, Juan Francisco. 2002. *Estado Ambiental Del Paraguay*. Asunción: SEAM—GTZ.

Fernández Jilberto, Alex E., and Barbara Hogenboom. 2010. Latin America and China—South-South Relations in a New Era. In *Latin America Facing China: South-South Relations Beyond the Washington Consensus*, ed. Alex E. Fernández Jilberto and Barbara Hogenboom. New York: Berghahn Books.

Finnemore, Martha, and Kathryn Sikkink. 1998. International Norm Dynamics and Political Change. *International Organization* 52 (4): 887–917.

FREPLATA. 2005. Análisis Diagnostico Transfronterizo Del Rio de La Plata Y Su Frente Marítimo, Documento Técnico. Montevideo. http://www.freplata.org/documentos/adt/default.asp.

Friedman, Elisabeth Jay, Kathryn Hochstetler, and Ann Marie Clark. 2001. Sovereign Limits and Regional Opportunities for Global Civil Society in Latin America. *Latin American Research Review* 36 (3): 7–35.

Fulquet, Gastón. 2015. ¿La Maldición de Los Recursos Naturales? Conocimiento Experto, Política Y Intereses Sectoriales En El Desarrollo de Biocombustibles En Sudamérica. *Brazilian Journal of International Relations* 4 (1): 39–70.

Green, Duncan. 1999. A Trip to the Market: The Impact of Neoliberalism in Latin America. In *Developments in Latin American Political Economy—States, Markets and Actors*, ed. Julia Buxton and Nicola Phillips. Manchester: Manchester University Press.

Grugel, Jean. 2001. *Democratization: A Critical Introduction.* Houndmills: Palgrave Macmillan.

Grugel, Jean. 2007. Latin America After the Third Wave. *Government and Opposition* 42 (2): 242–249.

Grugel, Jean. 2009. 'Basta de Realidades, Queremos Promesas': Democracy After the Washington Consensus. In *Governance after Neoliberalism in Latin America*, ed. Jean Grugel and Pía Riggirozzi. New York: Palgrave Macmillan.

Grugel, Jean, and Pía Riggirozzi. 2007. The Return of the State in Argentina. *International Affairs* 83 (1): 87–107.

Grugel, Jean, and Pía Riggirozzi. 2012. Post-Neoliberalism in Latin America: Rebuilding and Reclaiming the State After Crisis. *Development and Change* 43 (1): 1–21.

Grugel, Jean, and Jewellord Nem Singh. 2013. Citizenship, Democratisation and Resource Politics. In *Resource Governance and Developmental States in the Global South*, ed. Jewellord Nem Singh and France Bourgouin. Basingstoke: Palgrave Macmillan.

Gudynas, Eduardo. 2008. The New Bonfire of Vanities: Soybean Cultivation and Globalization in South America. *Development* 51 (4): 512–518.

Gudynas, Eduardo. 2009. Diez Tesis Urgentes Sobre El Nuevo Extractivismo. Contextos Y Demandas Bajo El Progresismo Sudamericano Actual. In *Extractivismo, Política Y Sociedad.* Quito: CAAP (Centro Andino de Acción Popular) and CLAES (Centro Latino Americano de Ecología Social).

Gudynas, Eduardo. 2010a. Agropecuaria Y Nuevo Extractivismo Bajo Los Gobiernos Progresistas de América Del Sur. *Territorios* (5): 37–54.

Gudynas, Eduardo. 2010b. Si Eres Tan Progresista ¿Por Qué Destruyes La Naturaleza? Neoextractivismo, Izquierda Y Alternativas. *Ecuador Debate, no.* 79: 61–81.

Gudynas, Eduardo. 2010c. La Ecología Política Del Progresismo Sudamericano: Los Límites Del Progreso Y La Renovacíon Verde de La Izquierda. *Sin Permiso* (8): 147–167.

Gutiérrez, Ofelia, and Daniel Panario. 2014. Implementación de Un Complejo Forestal Industrial, ¿ Una Política de Estado? Estudio de Caso: Uruguay. 7/2014. Documentos de Trabajo ENGOV. http://www.engov.eu/en/project-reports.html.

Gwynne, Robert N., and Cristobal Kay. 2000. Views from the Periphery: Futures of Neoliberalism in Latin America. *Third World Quarterly* 21 (1): 141–156.

Gwynne, Robert N, and Eduardo Silva. 1999. The Political Economy of Sustainable Development. In *Latin America Transformed—Globalization and Modernity*, ed. Robert N. Gwynne and Cristobal Kay. London: Arnold.

Hochstetler, Kathryn. 2002. After the Boomerang: Environmental Movements and Politics in the La Plata River Basin. *Global Environmental Politics* 2 (4): 35–57.

Hochstetler, Kathryn. 2003. Fading Green? Environmental Politics in the Mercosur Free Trade Agreement. *Latin American Politics and Society* 45 (4): 1–32.

Hochstetler, Kathryn. 2005. Race to the Middle: Environmental Politics in the Mercosur Free Trade Agreement. In *Handbook of Global Environmental Politics*, ed. Peter Dauvergne. Cheltenham: Edward Elgar.

Hochstetler, Kathryn. 2012. Democracy and the Environment in Latin America and Eastern Europe. In *Comparative Environmental Politics—Theory, Practice, and Prospects*, ed. Paul F. Steinberg and Stacy D. VanDeveer. London: MIT Press.

Hochstetler, Kathryn. 2013. South-South Trade and the Environment: A Brazilian Case Study. *Global Environmental Politics* 13 (1): 30–48.

Hochstetler, Kathryn, and Margaret E. Keck. 2007. *Greening Brazil—Environmental Activism in State and Society*. Durham and London: Duke University Press.

Hogenboom, Barbara. 2012. Depoliticized and Repoliticized Minerals in Latin America. *Journal of Developing Societies* 28 (2): 133–158.

Hogenboom, Barbara. 2015. New Elites around South America's Strategic Resources. In *Environmental Politics in Latin America—Elite Dynamics, the Left Tide and Sustainable Development*, ed. Benedicte Bull and Mariel Aguilar-Støen. Oxon: Routledge.

Hogenboom, Barbara, and Alex E Fernández Jilberto. 2009. The New Left and Mineral Politics: What's New? *European Review of Latin American and Caribbean Studies* (87): 93–102.

Høiby, Marte, and Joaquín Zenteno Hopp. 2015. Bolivia: Emerging and Traditional Elites and the Governance of the Soy Sector. In *Environmental Politics in Latin America—Elite Dynamics, the Left Tide and Sustainable Development*, ed. Benedicte Bull and Mariel Aguilar-Støen. Oxon: Routledge.

Kaltenthaler, Karl, and Frank O. Mora. 2002. Explaining Latin American Economic Integration: The Case of Mercosur. *Review of International Political Economy* 9 (1): 72–97.

Keck, Margaret E., and Kathryn Sikkink. 1998. *Activists Beyond Borders—Advocacy Networks in International Politics*. London: Cornell University Press.

Kirby, Peadar, and Barry Cannon. 2012. Globalization, Democratization and State-Civil Society Relations in Left-Led Latin America. In *Civil Society and the State in Left-Led Latin America—Challenges and Limitations to Democratization*, ed. Barry Cannon and Peadar Kirby. London: Zed Books.

Lapitz, Rocío, Gerardo Evia, and Eduardo Gudynas. 2004. *Soja Y Carne En El Mercosur - Comercio, Ambiente Y Desarrollo Agropecuario*. Montevideo: Coscoroba Ediciones.

Lewis, Tammy L. 2016. *Ecuador's Environmental Revolutions: Ecoimperialists, Ecodependents, and Ecoresisters*. Cambridge, MA: MIT Press.

Mumme, Stephen P., and Edward Korzetz. 1997. Democratization, Politics, and Environmental Reform. In *Latin American Environmental Policy in International Perspective*, ed. Gordon J. MacDonald, Daniel L. Nielson, and Marc A. Stern. Oxford: Westview Press.

Murray, Warwick E. 1999. Natural Resources, the Global Economy and Sustainability. In *Latin America Transformed—Globalization and Modernity*, ed. Robert N. Gwynne and Cristobal Kay. London: Arnold.

Newell, Peter. 2008. Trade and Biotechnology in Latin America: Democratization, Contestation and the Politics of Mobilization. *Journal of Agrarian Change* 8 (2 and 3): 345–376.

Newell, Peter. 2009. Bio-Hegemony: The Political Economy of Agricultural Biotechnology in Argentina. *Journal of Latin American Studies* 41 (1): 27–57.

O'Shaughnessy, Hugh, and Edgar Venerando Ruiz Díaz. 2009. *The Priest of Paraguay—Fernando Lugo and the Making of a Nation*. London: Zed Books.

Ortiz, María Selva, Javier Taks, Beatriz Schmid, and Stefan Thimmel (eds.). 2005. *Entre El Desierto Verde Y El País Productivo: El Modelo Forestal En Uruguay Y El Cono Sur*. Montevideo: Casa Bertolt Brecht and REDES-Amigos de la Tierra. http://www.redes.org.uy/wp-content/uploads/2008/10/entre-el-desierto-verde-y-el-pais-productivo.pdf.

Pakkasvirta, Jussi. 2010. *Fábricas de Celulosa*. Buenos Aires: La Colmena.

Panizza, Francisco. 2009. *Contemporary Latin America—Development and Democracy Beyond the Washington Consensus*. London: Zed Books.

Parker G., Cristián. 2015. Elite Views about Water and Energy Consumption in Mining in Argentina, Chile, Colombia and Ecuador. In *Environmental Politics in Latin America—Elite Dynamics, the Left Tide and Sustainable Development*, ed. Benedicte Bull and Mariel Aguilar-Støen. Oxon: Routledge.

Perreault, Tom. 2008. Popular Protest and Unpopular Policies: State Restructuring, Resource Conflict, and Social Justice in Bolivia. In *Environmental Justice in Latin America—Problems, Promise, and Practice*, ed. David V. Carruthers. London: MIT Press.

Peters, Ina. 2011. *Der Belo Monte Staudamm: Paradebeispiel Für Eine Erfolgreiche Zivilgesellschaft in Brasilien?* 9/2011. GIGA Focus Lateinamerika. Hamburg: German Institute of Global and Area Studies. http://www.giga-hamburg.de/de/publikationen/giga-focus/lateinamerika.

Phillips, Nicola. 1999. Global and Regional Linkages. In *Developments in Latin American Political Economy—States, Markets and Actors*, ed. Julia Buxton and Nicola Phillips. Manchester: Manchester University Press.

Phillips, Nicola, and Julia Buxton. 1999. Introduction. In *Developments in Latin American Political Economy—States, Markets and Actors*, ed. Julia Buxton and Nicola Phillips. Manchester: Manchester University Press.

Prevost, Gary. 2012. Argentina's Social Movements: Confrontation and Co-Optation. In *Social Movements and Leftist Governments in Latin America:*

Confrontation or Co-Optation?, ed. Gary Prevost, Carlos Oliva Campos, and Harry E. Vanden. London: Zed Books.

Prevost, Gary, Harry E. Vanden, and Carlos Oliva Campos. 2012. Introduction. In *Social Movements and Leftist Governments in Latin America: Confrontation or Co-Optation?*, ed. Gary Prevost, Carlos Oliva Campos, and Harry E Vanden. London: Zed Books.

Riggirozzi, Pía. 2009. After Neoliberalism in Argentina: Reasserting Nationalism in an Open Economy. In *Governance after Neoliberalism in Latin America*, ed. Jean Grugel and Pía Riggirozzi. New York: Palgrave Macmillan.

Riggirozzi, Pía, and Jean Grugel. 2009. Conclusion: Governance after Neoliberalism. In *Governance after Neoliberalism in Latin America*. New York: Palgrave Macmillan.

Rivera-Quiñones, Miguel A. 2014. Macroeconomic Governance in Post-Neoliberal Argentina and the Relentless Power of TNCs: The Case of the Soy Complex. In *Argentina Since the 2001 Crisis—Recovering the Past, Reclaiming the Future*, ed. Cara Levey, Daniel Ozarow, and Christopher Wylde. New York: Palgrave Macmillan.

Robinson, William I. 2008. *Latin America and Global Capitalism—A Critical Globalization Perspective*. Baltimore: The John Hopkins University Press.

Saguier, Marcelo. 2012a. Socio-Environmental Regionalism in South America: Tensions in New Development Models. In *The Rise of Post-Hegemonic Regionalism—The Case of Latin America*, ed. Pía Riggirozzi and Diana Tussie. Dordrecht: Springer.

Saguier, Marcelo. 2012b. La Integración Sudamericana Y Los Conflictos Socio-Ambientales. FLACSO Argentina. http://www.flacso.org.ar/actividad_vermas.php?id=1591.

Siegel, Karen M. 2016. Fulfilling Promises of More Substantive Democracy? Post-Neoliberalism and Natural Resource Governance in South America. *Development and Change* 47 (3): 495–516.

Silva, Eduardo. 2016. Afterword: From Sustainable Development to Environmental Governance. In *Environmental Governance in Latin America*, ed. Fabio de Castro, Barbara Hogenboom, and Michiel Baud. Basingstoke: Palgrave Macmillan.

Svampa, Maristella. 2012. Consenso de Los Commodities, Giro Ecoterritorial Y Pensamiento Crítico En América Latina. In *Observatorio Social de América Latina*, 15–37. Buenos Aires: CLACSO.

Taylor, Lucy. 1999. Market Forces and Moral Imperatives: The Professionalization of Social Activism in Latin America. In *Democracy without Borders—Transnationalization and Conditionality in New Democracies*, ed. Jean Grugel. London: Routledge.

Tedesco, Laura. 1999. NGOs and the Retreat of the State: The Hidden Dangers. In *Developments in Latin American Political Economy—States, Markets and*

Actors, ed. Julia Buxton, and Nicola Phillips. Manchester: Manchester University Press.

Thimmel, Stefan. 2010. Von Vázquez Zu Mujica: Bilanz Und Perspektiven. In *Uruguay—Ein Land in Bewegung*, ed. Stefan Thimmel, Theo Bruns, Gert Eisenbürger, and Britt Weyde. Berlin: Assoziation A.

Toni, Fabiano, Larissa C.L. Villarroel, and Bruno Taitson Bueno. 2015. State Governments and Forest Policy: A New Elite in the Brazilian Amazon? In *Environmental Politics in Latin America—Elite Dynamics, the Left Tide and Sustainable Development*, ed. Benedicte Bull and Mariel Aguilar-Støen. Oxon: Routledge.

Tussie, Diana, and Patricia Vásquez. 2000. Regional Integration and Building Blocks: The Case of Mercosur. In *The Environment and International Trade Negotiations—Developing Country Stakes*, ed. Diana Tussie. Houndmills: Macmillan Press.

Vanden, Harry E. 2012. The Landless Rural Workers' Movement and Their Waning Influence on Brazil's Workers' Party Government. In *Social Movements and Leftist Governments in Latin America: Confrontation or Co-Optation?*, ed. Gary Prevost, Carlos Oliva Campos, and Harry E. Vanden. London: Zed Books.

Vergara-Camus, Leandro. 2015. Sugarcane Ethanol: The Hen of the Golden Eggs? Agribusiness and the State in Lula's Brazil. In *Crisis and Contradiction: Marxist Perspectives on Latin America in the Global Economy*, ed. Susan J. Spronk and Jeffery R. Webber. Leiden: Brill Academic Publishers.

Wickstrom, Stefanie. 2008. Cultural Politics and the Essence of Life: Who Controls the Water? In *Environmental Justice in Latin America—Problems, Promise, and Practice*, ed. David V. Carruthers. London: MIT Press.

Zibechi, Raúl. 2010. Vorwärts in Die Vergangenheit – Der Linke Weg Zurück Zur Schweiz Lateinamerikas. In *Uruguay—Ein Land in Bewegung*, ed. Stefan Thimmel, Theo Bruns, Gert Eisenbürger, and Britt Weyde. Berlin: Assoziation A.

Zenteno Hopp, Joaquín, Eivind Hanche-Olsen, and Héctor Sejenovich. 2015. Argentina—Government-Agribusiness Elite Dynamics and Its Consequences for Environmental Governance. In *Environmental Politics in Latin America—Elite Dynamics, the Left Tide and Sustainable Development*, ed. Benedicte Bull and Mariel Aguilar-Støen. Oxon: Routledge.

CHAPTER 3

From "Open" to "Post-hegemonic" Regionalism

This chapter examines the evolution of regional cooperation in South America over the last 25 years in order to understand how regional *environmental* cooperation has come to develop to a large extent separately from regional organisations and in the margins of the regional integration processes promoted by governments. Throughout the time period examined in the book, South American governments have actively promoted regional integration through the creation of new regional organisations and by pursuing political and economic objectives at the regional level. Yet, these government-led regional integration processes and regional organisations have not become frameworks for robust forms of regional environmental cooperation. This chapter argues that this has been largely due to two related processes: the sidelining of environmental concerns in regional organisations by governments and the limited possibilities for participation of regional civil society networks working on environmental issues in regional organisations.

Reflecting changes in political priorities at the domestic level, the last 25 years have also seen a shift in the nature and logic of regional cooperation with the election of a wave of Leftist governments presenting a turning point. During the 1990s, South American governments followed the logic of "open" regionalism which was strongly promoted by the US and the EU. This reflected the priorities of the neoliberal reform agenda, so that the central focus of regional cooperation was economic liberalisation and free trade. Yet, the US and the EU approaches to regional integration also differed in important respects (Grugel 2004). Most importantly, the

© The Author(s) 2017   63
K.M. Siegel, *Regional Environmental Cooperation in South America*,
International Political Economy Series, DOI 10.1057/978-1-137-55874-9_3

EU also sought to promote its own model of regional integration in other parts of the world which includes not only economic integration, but also political cooperation on a range of other issues. Various European actors have thus actively engaged in attempts to diffuse European norms of democracy, social inclusion, environment and sustainability and regional cooperation beyond Europe in an effort to foster a distinctive model of regionalism based on the European model. Initially, this also seemed to promote regional environmental cooperation in the Southern Cone. With European support, the regional organisation Mercosur, created by Argentina, Brazil, Paraguay and Uruguay in 1991, promised to become an important framework for addressing regional environmental concerns. Yet, over time it has become clear that there is little political will to strengthen Mercosur and follow the EU's path of building supranational institutions and spillover from economic integration to other policy areas has remained much more limited than in the European case. At the same time, although there is a relatively high level of institutionalisation of environmental cooperation in the Mercosur framework, the agendas of the relevant institutions have changed frequently while their resources have been restricted, dialogue with regional civil society networks working on socio-environmental concerns has been limited, and Mercosur has not been given the mandate to deal with some of the region's most important shared and transboundary socio-environmental concerns. As a consequence, the relevance of Mercosur for environmental cooperation in the region has declined over time.

With the election of a wave of Leftist governments across the region in the 2000s, the nature of regional cooperation changed in important ways as social objectives and inclusion became more important while trade as the main driver for regional integration and the EU model lost in significance. Mirroring changes at the national level, the "post-hegemonic" or "post-neoliberal" (Riggirozzi and Tussie 2012) approach towards regional cooperation promoted by progressive South American governments in the twenty-first century sought to defy neo-liberalism and reassert the region's autonomy vis-à-vis the US while focussing on important regional and domestic concerns. To achieve these aims, two new regional organisations, Unasur and Alba, as well as the Initiative for the Integration of Regional Infrastructure in South America (IIRSA) were launched. The latter is a large-scale South America-wide initiative aimed at improving the transport, energy and communications infrastructure in order to open up trade routes within South America and improve the physical integration within the region. However, the initiative has also been used to facilitate commodity exports and therefore contributed to the regionalisation of agribusiness in the Southern Cone. Neo-extractivism has thus clearly moved into the regional sphere and affects regional politics and cooperation. This also means that the socio-environmental dimensions of resource governance are often transboundary and have implications for regional cooperation. This is outlined in the second part of this chapter examining IIRSA and the expansion of agribusiness in Paraguay, a country that is often overlooked in studies of regional cooperation in South America. Both examples demonstrate the ad hoc and inconsistent nature of government responses to challenges arising from the regional and socio-environmental dimensions of resource governance. They also uncover the continuation of a trend that has already become evident in the evolution of Mercosur in terms of sidelining environmental concerns and civil society networks working on them in regional organisations. As a result, the various regional organisations and integration processes led by governments have not become frameworks for robust forms of regional environmental cooperation.

THE DECLINING RELEVANCE OF MERCOSUR FOR REGIONAL ENVIRONMENTAL COOPERATION

Following the debt crisis of the 1980s, regionalism took a new turn in South America in the 1990s. In the context of the return to democracy and the Washington Consensus, the Andean Community, a regional organisation that had existed since 1969, was relaunched while the Southern Cone countries of Argentina, Brazil, Paraguay and Uruguay created the Common Market of the South, or Mercosur, in 1991. These regional projects were part of the "new" or "open" regionalism and locked in the neoliberal reform agenda at a regional level (Riggirozzi 2012a, 32). At the same time as these regional organisations within South America developed, the US sought to promote its interests in the region by promoting the idea of a Free Trade Area of the Americas (FTAA). Although economic objectives were clearly predominant during the 1990s, they were nevertheless not the only objectives of regional cooperation. Regional cooperation on issues other than trade, including the environment and sustainability, has been promoted strongly by the EU, and European agencies have provided substantial aid to Mercosur in an attempt to project the EU's model of regional integration in other parts of the world

(Grugel 2004; Lenz 2012, 161–162). Yet, over time disagreements between European visions of regional integration in the Southern Cone and those of the Southern Cone countries have become evident. It has thus become clear that there is little political will to strengthen Mercosur and create supranational institutions similar to those of the EU and spillover from economic integration to other policy areas has remained much more limited than in the European case. Instead, the regional organisation has been kept alive by the direct intervention of presidents (Malamud 2015). The evolution of Mercosur is particularly interesting because initially it seemed like a promising framework for regional environmental cooperation. However, although there is a relatively high level of institutionalisation of environmental cooperation in the Mercosur framework, the agendas of the relevant institutions have changed frequently while their resources have been restricted, dialogue with regional civil society networks working on socio-environmental concerns has been limited, and Mercosur has not been given the mandate to deal with some of the region's most important socio-environmental concerns. As a consequence, the relevance of Mercosur for environmental cooperation in the region has declined over time.

Although Mercosur was created mainly as a trade agreement focussing on economic objectives as its name *Mercado Común del Súr* (Common Market of the South) also indicates, environmental concerns have been on Mercosur's agenda since its creation with the founding treaty referring to the preservation of the environment in its preamble (Hochstetler 2003, 5–6; 2005, 351). A year after Mercosur's creation, the Specialised Meeting on the Environment (*Reunión Especializada de Medio Ambiente* or REMA) was created. This was later upgraded to the working subgroup 6 on the environment (*Subgrupo de Trabajo 6 Medio Ambiente* or SGT6) (Laciar 2003, 51–61; Secretaría Administrativa del MERCOSUR 2002, 16–17; Torres and Diaz 2011, 205–206; Tussie and Vásquez 2000, 196–197; UNEP and CLAES 2008, 105). In the Mercosur framework, several such working subgroups (SGTs) exist for different topics. The SGT6 consists of government officials from the ministries of environment or equivalent from the four member states and meets on a regular basis. In addition to the SGT6, a first meeting of environment ministers of the four Mercosur countries took place in 1995 and was institutionalised in 2003 with the creation of the Meeting of the Environment Ministers (*Reunión de Ministros de Medio Ambiente*) taking place about twice a year (Moreno 2011, 70; UNEP and CLAES 2008, 106–107). Taken together, these two Mercosur institutions provide the basis for a relatively high level of institutionalisation of environmental concerns in the regional organisation. In terms of policy-making, the Mercosur countries have adopted several joint strategies or declarations on different environmental topics such as climate change, desertification, biodiversity or sustainable production and consumption, and after several years of negotiations, the Mercosur countries approved an environmental framework agreement, the *Acuerdo Marco sobre Medio Ambiente del Mercosur*, in 2001 (Laciar 2003, 89–153). The fact that an environmental framework agreement was approved within the first decade of Mercosur's existence can be seen as an achievement because it demonstrates the efforts the member states made towards the coordination of their environmental policies (Torres and Diaz 2011, 209).

European support has played an important role in promoting regional environmental cooperation in the Mercosur framework, but it was also shaped by global processes on environmental matters and trade-related considerations. By developing various declarations and joint strategies, the Southern Cone governments have used Mercosur to demonstrate a commitment to international environmental processes. The creation of Mercosur coincided with the preparations for the Rio Summit a year later. The presidents of the Southern Cone approved the Canela Declaration prior to the Rio Summit in 1992 with the aim of examining the topics of the conference. Although the declaration is not officially a Mercosur document as it also includes Chile, the fact that it was approved by the presidents of all four Mercosur countries made it a reference document for the integration process that had just started and it has been considered a first joint action of Mercosur on the topic of the environment (Laciar 2003, 48–49; Moreno 2011, 70; Torres and Diaz 2011, 204). Since then, Mercosur declarations and strategies on environmental topics have often related to global conventions and reiterated the objectives and commitments of these (Mercosur 2008, 22; Torres and Diaz 2011, 224; UNEP and CLAES 2008, 107). Yet, Mercosur as a regional organisation has had very little visibility and impact on these global processes. In terms of exercising influence or even stating joint positions, individual countries, in particular the bigger states Brazil and Argentina, and other groupings such as the whole Latin America and Caribbean region, or the G77, a coalition of over 130 countries of the South, have been more prominent than Mercosur (Mercosur 2008, 9). Moreover, since the Copenhagen Summit in 2009, Brazil has increasingly coordinated its position in the climate change negotiations with other emerging powers, such as China, India and South Africa rather than with its Mercosur partners (Hochstetler 2012; Hochstetler and Viola 2012).

Furthermore, the expectation that trade and environmental policy are linked has also contributed to the inclusion of environmental concerns on the Mercosur agenda. When Mercosur was created, the link between the environment and trade was very much debated in the context of the negotiations on the General Agreement on Tariffs and Trade and the World Trade Organization as well as the discussions on the North American Free Trade Agreement (NAFTA) (Laciar 2003, 27–30; Torres and Diaz 2011, 203; UNEP and CLAES 2008, 104). However, within the Mercosur framework, the emphasis of governments has clearly been on trade and competitiveness rather than environmental protection. The Southern Cone governments were thus very concerned that environmental regulations could be turned into non-tariff barriers or that environmental concerns may limit important economic activities. As a result, the emphasis in the work given to the REMA and later the SGT6 was clearly on trade-related aspects and in particular the concern that environmental regulations could be used as non-tariff barriers in the region. Moreover, environmental topics with sensitive implications for trade have frequently been withdrawn from the agenda of Mercosur's environmental forums and addressed elsewhere without the involvement of environmental experts (Devia 1998, 30–34; Hochstetler 2003, 14; Laciar 2003, 61–78; Mercosur 2008, 21; Torres and Diaz 2011, 206). A declaration issued by the Mercosur environment ministers prior to the environmental summit held in Rio de Janeiro in 2012 reaffirmed this stance by stressing the need to avoid "green protectionism" at the global level and obstacles to trade because of environmental concerns (Mercosur 2012).

The prioritisation of other concerns is also clear when looking at the way Mercosur's environmental forums have functioned in practice and the significant constraints they have been subjected to. Several accounts, for example, point out that a very ambitious and more detailed initial proposal for the environmental framework agreement was watered down considerably before the Mercosur governments agreed to approve a pragmatic final version (Hochstetler 2005, 352–353, 2003, 17–23; Laciar 2003, 89–153; Moreno 2011, 71; UNEP and CLAES 2008, 106). The lack of concrete indications of how the framework agreement would be implemented further demonstrates its limitations and further negotiations would be necessary to ensure implementation (Torres and Diaz 2011, 209). Furthermore, although the SGT6 has been important in terms of bringing together government officials working on environmental issues from all the Mercosur countries in regular meetings, it has been limited in what it could achieve because the SGT6 does not have resources of its own and cannot set its own agenda. Instead, it depends very much on external donors and Mercosur's main decision-making bodies which are dominated by the ministries of foreign affairs and economy. The SGT6 may propose additional agenda items, but overall it acts mostly as a technical advisory committee responding to questions and tasks set by Mercosur's main decision-making bodies (Hochstetler 2003, 4–6, 2005, 351; Laciar 2003, 34–36; UNEP and CLAES 2008, 23–26). Overall, there is little political will to provide more resources or autonomy to the environmental institutions of Mercosur, and these are weak both in comparison with other Mercosur institutions and environmental bodies of other regional organisations such as NAFTA or the EU (Hochstetler 2003, 12–13).

The Mercosur Parliament, or Parlasur, is another actor that has repeatedly argued in favour of giving environmental concerns more attention, but has had only limited influence. In 2010, the chairman of the Brazilian delegation to the Parliament pointed out that 30–40% of proposals of the Mercosur Parliament relate to the environment (Câmara dos Deputados Brazil 2010a). The Parliament has also made suggestions, such as creating a transboundary area under environmental protection in the triple-border area between Argentina, Brazil and Paraguay, to protect the biodiversity of this region (Câmara dos Deputados Brazil 2010b; Torres and Diaz 2011, 215–216), and it has worked very actively on the Guaraní aquifer which the four original Mercosur countries share. However, the powers and competences of the Mercosur Parliament are extremely limited and do not include setting rules (Malamud 2015, 170–171). As Malamud and Dri (2013, 234) point out, the Southern Cone executives have been reluctant to empower an institution that could challenge their own power. On the whole, this means that those Mercosur institutions that have the most interest in promoting environmental cooperation do not have much power within the Mercosur structures. This is a clear indicator of the lack of political will on the part of the Southern Cone governments to make Mercosur an effective framework for addressing regional environmental concerns.

This is also evident in the trend that has become apparent over the years which demonstrates that Mercosur's environmental forums have not been given the mandate to address many of the region's most important shared and transboundary socio-environmental concerns. This started during

Mercosur's first decade and has continued to this day. For example, one of the major obstacles in the development of Mercosur's environmental framework agreement was Argentina's objection to including biosafety issues. In contrast to the other Mercosur countries which initially restricted the use of GM technologies, Argentinean policy-makers had embraced them during the 1990s. Although the Argentinean delegate of the SGT6 had widely consulted in Argentina regarding a first, more detailed Mercosur environment protocol, this was rejected by Argentina's delegation in 1997 with the inclusion of biosafety issues being one of the main concerns. The agreement was only approved after the topic of biosafety, an important environmental issue in all the Mercosur countries which has been the subject of criticism by environmentalists and small farmers, had been excluded (Hochstetler 2003, 20; Newell 2008, 363). The SGT6 has also been sidelined or excluded from discussions on other environmental issues such as the plans for the construction of the hidrovía in the La Plata basin or the GEF project on the Guaraní aquifer (Hochstetler 2003, 13–14; 2011, 137; Mercosur 2008, 18; Moreno 2011, 73; Villar and Ribeiro 2011, 651), both discussed in more detail in the following chapter. Finally, Mercosur's lack of involvement has been particularly striking in relation to the pulp mill dispute between Argentina and Uruguay which broke out over the use of a shared river in the 2000s and which escalated to an international conflict. Yet, the governments in the region, including Brazil, decided not to address the conflict within the Mercosur framework, and it was eventually ended in 2010 through a ruling of the International Court of Justice (ICJ) in The Hague far away from the region and without the involvement of existing regional institutions (Chidiak 2012).

Moreover, there is often very little coordination between the SGT6 and other Mercosur forums which deal with topics that also touch on environmental issues (Mercosur 2008, 6). One interviewee also noted that even the declarations or strategies on environmental topics that Mercosur has adopted are often not taken into account later on and follow-up guidance in terms of how they are to be implemented is frequently lacking:

> That's the big problem I see in the environmental policies in Mercosur, the later policies [on other topics] don't take the previous ones [on environmental issues] into account.[1]

The impression of Mercosur's lack of relevance for environmental cooperation in the Southern Cone is further reinforced when examining the role of civil society in the regional organisation. Initially, some environmental organisations expressed some interest in the regional organisation. The Argentinean organisation *Fundación Ambiente y Recursos Naturales* (FARN), for example, published several analyses on the environmental aspects of free trade during Mercosur's first decade, and the *Centro Latinoamericano de Ecología Social* (CLAES), based in Montevideo, has included regional integration and trade as one of its thematic areas (CLAES 2013; FARN 2013). However, interest on the part of civil society organisations has declined over time as it became clear that access to decision-making was difficult to get and that those environmental institutions which provided a minimum of access did not deal with many of the most important regional environmental problems. Consequently, civil society participation in SGT6 meetings has gone down consistently over the years (Hochstetler 2011, 138). Yet, various other organisations involving civil society have continued to borrow the name of Mercosur and thus build on the identity that the regional organisation provides, but with few links to the government-led cooperation processes and regional organisations. This includes Mercociudades, a network of cities that was created in 1995 with the aim of promoting the participation of municipalities in the process of regional integration. Another example is the Centre of Education for Regional Integration (*Centro de Formación para la Integración Regional* or CEFIR), based in Montevideo. The centre acts as the technical secretariat of *Somos Mercosur*, an initiative which aims at strengthening civil society participation in Mercosur and involving the population in the regional integration process (Somos Mercosur 2013; Torres and Diaz 2011, 213–215). The agenda of the initiative includes environmental issues and, like Mercociudades, goes beyond the official Mercosur agenda.

These initiatives clearly demonstrate the interest of some civil society groups to engage in regional cooperation and address shared environmental concerns from a regional perspective. However, they also demonstrate ambiguity on the part of governments and in particular the progressive governments of the left tide. On the one hand, these have spoken out in favour of more civil society participation in general and in regional cooperation, and they have taken some steps towards this (Briceño Ruiz 2012). The *Somos Mercosur* initiative, for example, was launched by the Uruguayan government during the presidency of Tabaré Vázquez (Briceño Ruiz 2012, 181–182; Serbin 2012, 156). On the other hand, these new initiatives have not been formally integrated into the Mercosur structure and institutionalised channels for civil society participation in regional cooperation remain weakly developed, particularly in relation to environmental concerns which are often linked to the highly sensitive topic of natural resource governance.

Finally, while there have been some important achievements of externally funded projects, disagreements between Mercosur policy-makers and European donors have also become evident. As the Southern Cone governments do not provide much funding for Mercosur's environmental forums, these heavily depend on external funding to be able to address environmental issues (Fulquet 2010, 15; Mercosur 2008, 2, 16). This is even the case for attending meetings with Paraguay, the poorest of the Mercosur countries, hardly being able to send representatives to all the meetings (Hochstetler 2011, 140). Environmental cooperation in Mercosur has received support from different donors, but the largest projects have been funded by the German cooperation agency (*Deutsche Gesellschaft für Technische Zusammenarbeit* or GTZ[2]) and the European Commission. The project "Competitiveness and Environment" ("Competitividad y Medio Ambiente" or CyMA according to its Spanish acronym) was funded by the GTZ and ran for five years from 2002 to 2007. The overall aim of the project was to elaborate and implement a strategy to increase the competitiveness and environmental efficiency particularly of small and medium enterprises. Two years after the end of the CyMA project, Mercosur made an agreement with the EU for another project which would amongst other things continue cooperation on the topic of sustainable consumption and production.

The CyMA project was important to collect information from the four Mercosur countries to get an overview of activities and projects in relation to clean production in each country. One interviewee stated that the joint work on the topic of sustainable production was also essential to establish a common basis:

> When we started to talk about that...some were talking about clean production, others were talking about certification, about ISO, the languages were completely different. It was like the tower of Babel, the four countries were talking about different things. When the project finished, at least we all had a clear concept for each country...[3]

Moreover, the project led to the approval of a declaration of principles of clean production by the ministers of the environment in 2003. Four years later, this was strengthened by a policy on the promotion and cooperation on sustainable production and consumption in Mercosur (Mercosur 2003, 2007a). The CyMA project also contributed to the strengthening of several Mercosur forums. It reinforced in particular the SGT6 and compensated for the lack of resources the SGT6 was suffering from. Moreover, the project encouraged cooperation between different Mercosur working groups, in particular by involving the SGT7, the working subgroup on industry, in the implementation of the project. As a result, the topic of the project also gained a more horizontal relevance, rather than being seen only as an environmental matter. The project also supported the development of the Mercosur environmental information system. Overall, the CyMA project thus led to a significant strengthening of the environmental agenda in Mercosur and temporarily deterred the marginalisation of Mercosur's environmental forums and environmental policies. It also contributed to the reduction of asymmetries in Mercosur by ensuring that representatives from all countries were able to participate in the meetings on a regular basis and at an equal level. In particular, representatives from the smaller member states appreciated the fact that the project reduced the gap to the bigger member states in terms of level of technical capacity and the capacity to contribute to the discussions at the regional level in a proactive way and on a regular basis (Mercosur 2007b, 13; Monge and Jacoby 2007).

However, it is significant that the objective of European donors was not primarily to promote a particular environmental topic or solve a specific regional environmental problem. Rather, its aim was to promote EU-style regional integration in the form of Mercosur and combine this with European norms of environmental protection and sustainable development. By linking support for Mercosur to assistance for environmental concerns, European agencies thus play an important role in terms of promoting Mercosur as a framework for regional environmental cooperation. EU policy documents have frequently reiterated the importance of the EU model for Mercosur and how Mercosur could learn from the EU (European Commission 2007, 24–36). Similarly, in the great majority of project publications, the prologue by the German cooperation agency states that Germany as well as the EU, based on their own experience, believe in regional integration as a solution to environmental and other global problems and therefore follow the development of Mercosur with great interest (Mercosur 2004a, 12, b, 12, 2007b, 7). However, this comparison of Mercosur with the EU has led to very specific expectations, and as the evaluation of the CyMA project shows, this has also led to some friction:

> Particularly from a perspective that applies, misleadingly, the supranational institutional model of the EU as a reference model, expectations that are too high with regards to the objectives of regional integration and which do not conform to Mercosur's own model of integration tend to be generated. As a result, the complexity of the decision-making processes of Mercosur, as well as the added value of decisions taken at the regional level are occasionally underestimated which obstructs a just and unanimous appreciation of strategic products, like for example the elaboration and approval of a regional policy on sustainable production and consumption. (Monge and Jacoby 2007, 51, author's translation)

The evaluation thus recommends that in any future cooperation projects with Mercosur, the partners need to be aware of the structures of Mercosur because only if they realise the potential as well as the limitations of Mercosur's integration model, will it be possible to adjust the expectations to reality (Monge and Jacoby 2007, 71).

The example clearly demonstrates that there is friction in particular over the model of regional integration. Although Mercosur policy-makers initially adopted the EU model enthusiastically, there were no detailed plans of how to achieve this (Lenz 2012, 161–162). Two decades later, it has become very clear that there is little political will to strengthen Mercosur and follow the EU's path of building supranational institutions. European donors have struggled to adapt to this reality and have continued to insist on the relevance of the European model. While European donors have very much promoted Mercosur as a channel for regional environmental cooperation, the Southern Cone governments resist this on two levels. First, they have not strengthened Mercosur's institutions in general, so that to some extent the overall weakness of Mercosur's environmental institutions is also a reflection of the low level of institutionalisation in Mercosur generally (Hochstetler 2011, 136). In addition, environmental protection is not a priority for the Mercosur governments (Mercosur 2008, 2; Tussie and Vásquez 2000, 199), and this is reflected in the institutions of the regional organisation. Consequently, the Mercosur governments have resisted the diffusion of environmental norms, and environmental cooperation in Mercosur remains very fragmented. This echoes other studies which have found that in the area of democracy promotion and in particular social citizenship as an element of democracy, Mercosur's governing elites have equally resisted the adoption of norms promoted by the EU, suggesting that norm diffusion is an extremely ambitious goal (Grugel 2007).

Given these discrepancies between the priorities of donors and recipients, it is perhaps not surprising that regional environmental cooperation in the Mercosur framework suffers from interruptions and discontinuity. The perspective for cooperation in practice is therefore very much limited by the duration of specific projects. In the case of the CyMA project, the GTZ decided not to fund a second implementation phase as was hoped on the Mercosur side and that meant that the implementation of the policies on sustainable production and consumption was not ensured. Several years after the end of the project, a former government official still listed the regular meetings and the approval of the policies as important achievements, but also pointed to the lack of continuity after the project finished:

> What the CyMA project achieved is that Mercosur through the ad hoc group and the SGT6 had a permanent organised activity for about five years on the topic of clean production. That was a very important aspect. From this came some guidelines on cleaner production and also the declaration on cleaner production of Mercosur...All these issues were very much encouraged by this project. And the technical groups that support these topics were able to interact and to meet at the regional level thanks to this project...In my opinion what is left is very meagre. It should be used more. Very little continued afterwards...[4]

In addition, the interviewee commented that a few years are nothing to address a topic like sustainable development and that a much longer time frame is needed to do this successfully. However, while the work with the German cooperation agency had been very good, the GTZ had now closed its office in Buenos Aires. Similarly, another interviewee noted that the ad hoc group on Competitiveness and Environment created during the project had not been active after the end of the project and that there was hardly any follow-up of the activities.[5]

On the whole, there is thus a relatively high level of institutionalisation of regional environmental cooperation within the Mercosur framework, and in fact, Mercosur's environmental institutions are the most developed in the region as far as government-led cooperation is concerned. However, this has not translated into effective cooperation in practice. Mercosur's environmental forums have very little autonomy and are in a weak position vis-à-vis other institutions and interests. In this context, joint environmental strategies or declarations frequently lack implementation and are often not taken into account in other policies developed later. Moreover, the Southern Cone governments have repeatedly decided not to give Mercosur the mandate to address several of the region's most important transboundary environmental concerns. In addition, the projects that have been relatively successful are highly donor-dependent, and the relatively short time frames of external funding have been a serious limitation. Moreover, disagreements with European donors whose main objective is to promote their own model of regional integration have also become evident. At the same time, several regional civil society initiatives continue to build on a shared Mercosur identity. This clearly demonstrates an interest on the part of some civil society groups to address environmental concerns from a regional perspective, but also the reluctance of governments to integrate these endogenous drivers into the formal structures which would strengthen regional environmental cooperation. Altogether this suggests that regional environmental cooperation in the Mercosur framework is driven less by a desire to solve particular regional environmental problems than by the desire to make symbolic commitments and replicate EU-style regional integration. It is not a surprise then that a document on the evolution and perspectives of environmental issues within Mercosur put together during the 9th Meeting of the Mercosur Environment Ministers in November 2008 refers to the "progressive draining of Mercosur's environmental agenda" (Mercosur 2008: 2, author's translation). Although Mercosur has often been described as the most developed regional organisation outside Europe (Gardini 2011, 235; Kaltenthaler and Mora 2002, 73; Telò 2006, 131), its relevance for environmental cooperation in the Southern Cone has, in fact, decreased over the last two decades.

NEO-EXTRACTIVISM MOVING INTO THE REGIONAL SPHERE

At the same time as Mercosur's stagnation and lack of emulation of the EU model became evident (Malamud 2005), the nature of regional cooperation in South America also changed in several important respects. Most importantly, there have been notable changes in the content and objectives of regional cooperation as well as the rhetoric of political leaders. Reflecting developments at the national level and benefitting from the increased political and economic autonomy offered by the commodity boom, left tide leaders have also stressed the importance of domestic social objectives and autonomy from external pressure, and in particular US influence, in regional cooperation. To this end, new regional organisations, the Venezuelan-led Alba and the Brazilian-led Unasur, as well as the Community of Latin American and Caribbean States (CELAC) bringing together the sovereign countries of the Americas with the exception of the US and Canada, were launched while the US-led FTAA collapsed in 2005. The new regional organisations shifted the focus from economic and market priorities to social concerns, notably education, health and employment. Important elements of the new approach to regional cooperation also include cooperation on regional needs such as energy and infrastructure as well as the establishment of an alternative regional development bank in the form of the Banco del Sur which has increased the autonomy from traditional financial institutions (Chodor 2015, 147–161; Chodor and McCarthy-Jones 2013; Luchetti 2015; Riggirozzi 2012b). A shift in priorities is also evident in old regional organisations, notably Mercosur, where gradually political objectives such as labour rights or a structural convergence fund to address some of the asymmetries within Mercosur gained ground (Briceño Ruiz 2012; Riggirozzi 2012b, 430; Serbin 2012).

South American regional cooperation in the new millennium contrasts in many ways with regional cooperation a decade earlier which focussed much more on trade liberalisation and has often been seen as a response to external circumstances, either in the form of US pressure or as a way of strengthening the region's position in the face of economic globalisation (Chodor and McCarthy-Jones 2013, 214; Grugel and Hout 1999). Consequently, some scholars have noted that a new era of "post-neoliberal", "post-hegemonic" and "post-trade" regional cooperation has begun (Riggirozzi 2012b; Riggirozzi and Tussie 2012). This does not imply a complete break with the past, but rather that the focus of regional cooperation has shifted. As Riggirozzi and Tussie (2012, 6) point out, even if the new regional projects are still at an early stage with a very low level of institutionalisation, they are important because they have changed the parameters of regional cooperation and created new conceptions of "what regionalism *is* and *is for*" or, as Chodor (2015, 147) puts it, they have laid "the basis for a new regional common sense" with distinctly different priorities from the neoliberal approach of the 1990s. In parallel to these developments, regional cooperation has seen a shift in the scale of cooperation from distinct regions within South America, such as the Southern Cone or the Andean region, towards cooperation at a larger South American or even Latin American-wide scale. As Emerson (2014) argues, this has gone hand in hand with a new production of "Latin Americaness" which has been evident particularly in the case of Alba.

With the shift in scale and the new emphasis on domestic and regional needs, natural resource governance has become a central element of regional integration. Venezuela in particular used its oil resources to strengthen regional institutions and create bilateral programmes within the region in order to increase the region's autonomy and challenge the position of the US (Chodor 2015, 155; Hogenboom 2015, 125; Lewis 2016, 168–169). In addition, at the start of the millennium, IIRSA was launched under Brazilian leadership. It was formally incorporated into Unasur in 2010 and is now part of the South American Council of Infrastructure and Planning (COSIPLAN). The objective of this large-scale South America-wide initiative is to improve the transport, energy and communications infrastructure in order to improve the physical integration between South American countries. With over 500 planned projects, the initiative intends to build stronger links between South America's largest cities and economic hubs. This requires overcoming significant distances and considerable geographical barriers like the Andes or the Amazon basin and therefore also affects many of the region's most remote areas (Carciofi 2012; Garzón and Schilling-Vacaflor 2012; Saguier 2012a, 130; Scholvin and Malamud 2014). Opening up trade routes is important for the economic integration of a region where intra-regional trade has historically been very low compared to other regions (Hochstetler 2011, 140–141), although studies have noted that IIRSA has mostly been used to promote individual projects of national interest rather than regional integration (Scholvin and Malamud 2014, 5). Several projects of the initiative have played an important role in terms of facilitating commodity exports and have, for example, contributed to the regionalisation of agribusiness in the Southern Cone discussed below. Over the last years, natural resource governance has also become a focal area of the Economic Commission of Latin America and the Caribbean (ECLAC) which plays a crucial role in regional policy-making through data gathering and analysis as well as policy recommendations. Recent ECLAC reports have highlighted the need to use mineral rents for the public sector and improve the governance of mining, oil, hydropower and water. Socio-environmental conflicts in relation to these areas have been mentioned only briefly, but have not been examined in depth (Hogenboom 2015, 126).

Neo-extractivism has thus clearly moved into the regional sphere and shapes regional politics and cooperation, or as Saguier (2012a, 126) argues, regional cooperation in South America in the twenty-first century has become "resource-driven". This also means that the socio-environmental dimensions of resource governance are often transboundary and have implications for regional cooperation. Looking at IIRSA and the expansion of agribusiness in Paraguay, a country that is often overlooked in studies of regional cooperation in South America, the final part of the chapter highlights the ad hoc and inconsistent nature of government responses to challenges arising from the regional socio-environmental dimensions of resource governance. This demonstrates the continuation of a trend that has already become evident in the evolution of Mercosur in terms of sidelining environmental concerns in regional organisations and the lack of possibilities for civil society participation. As a result, the various regional organisations and integration processes led by governments have not become frameworks for robust forms of regional environmental cooperation.

Contestations over how natural resources should be managed, who should take decisions and who should benefit, have made IIRSA the only regional initiative in South America facing a concerted opposition movement (Hochstetler 2011, 144–145). Many infrastructure projects have significant consequences for the livelihoods of people in the affected areas as well as for the physical environment. Although some assessments of environmental and social impact have been carried out, these have come late and have been applied inconsistently, thus not meeting the ambitious targets that had been set (Hochstetler 2011, 143–144, 2013, 43–44). As a consequence, IIRSA has faced large-scale regional opposition of civil society groups and affected communities (Garzón and Schilling-Vacaflor 2012; Hochstetler 2011, 144; Phillips and Cabitza 2011; Saguier 2012a, 134–135). Furthermore, Brazil's role in promoting the initiative has been contested in neighbouring countries fuelled by the perception that the new infrastructure mostly benefits South America's largest country (Burges 2005, 451; Malamud 2012, 175; Phillips and Cabitza 2011). Other reasons for protests include the view that the infrastructure developments would largely favour agribusiness (Hochstetler 2011, 145). These debates have raised concerns that IIRSA is more a tool for interconnecting global markets benefitting transnational capital involved in the export of resources than for regional integration strengthening the internal market and supporting the economic development of the region (de Geus 2011).

The expansion of agricultural production for export and soybean in particular has led to the physical and economic integration and spatial reorganisation of the Southern Cone region. This can be seen as a different form of regionalism which has not been planned by governments, even though governments have facilitated the process (Giraudo 2014). Turzi (2011) outlines how the creation of a single regionalised soybean chain has turned the Southern Cone into a "soybean republic". As agribusiness relies on large-scale and large-volume production together with technology to increase efficiency and secure competitive advantages, large parts of Argentina, Bolivia, Brazil, Paraguay and Uruguay have become closely integrated into a network of production, processing and distribution. Having benefitted from the market liberalisation of the 1990s, the agribusiness sector has pushed for and benefitted from new regional infrastructure developments. As a result of this process, national boundaries are losing in significance, but this new form of regional integration has been shaped largely by the needs and interests of the transnational corporations and companies driving the agribusiness sector. Not surprisingly, the regionalisation of agribusiness in the Southern Cone has also had domestic and regional political consequences which have impacted on power relations between and within countries and shaped regional cooperation. This has become particularly evident in the controversial impeachment process, perceived by many as a "constitutional coup", against the former president of Paraguay, Fernando Lugo, in June 2012.

Lugo came to power in 2008 with an ideological programme promising to address the country's high levels of poverty and inequality, tackling corruption, land reform and protecting the rights of indigenous people (Lambert 2011a, 76, 2011b, 177; Lugo Méndez 2013; O'Shaughnessy and Ruiz Díaz 2009). As Lambert (2011b, 184–185) notes, these were challenging objectives in a country where corruption was ingrained—in 2002, Paraguay was ranked as the third most corrupt country worldwide—and where elite interests were well entrenched. Moreover, inequality is high in Paraguay, even compared to other Latin American countries. This is particularly evident in relation to land distribution with 1% of landowners in possession of 77% of arable land. Furthermore, Paraguay's tax system is characterised by the absence of a direct income tax, and no effective export tax on agricultural products, using instead income from regressive and indirect taxation, mostly in the form of value-added tax (Lambert 2011b, 184–185). Significantly, access to land in Paraguay goes hand in hand with access to power and wealth with almost all members of Congress being closely linked to the small landowning elite (Lambert and Nickson 2013, 453). The 2008 election ended more than half a century of conservative rule by the Colorado Party and constituted the first time in the country's history that power was handed over to an opposition party peacefully and on the basis of elections (Lambert 2011b, 177–178). Despite Lugo's high personal popularity, his government from the start depended on a fragmented and divided coalition which did not have an absolute majority in a Congress with extensive constitutional powers. In this context, Lugo managed to make some important advances, in particular in expanding public health care and establishing a system of conditional cash transfer based on the example of the Brazilian *Bolsa Familia* programme, but his initiatives of land and tax reform were frequently blocked (Lambert 2011b, 181–182; Lambert and Nickson 2013; Wachendorfer 2013).

Lugo's presidency came to a premature end in June 2012 when the Congress used its powers to impeach him almost overnight in an extraordinarily rapid process. The event that was used as a reason for the impeachment was a clash between landless peasants and police in the area of Curuguaty in which eleven peasants and six policemen lost their lives. The exact circumstances of the massacre have remained obscure (Wachendorfer 2013, 1), but the powerful opposition cited "poor performance" and Lugo's inability to deal with growing insecurity to get rid of a president whose commitment to reform threatened their interests (Lambert 2012; Lambert and Nickson 2013). After an interim government, elections in April 2013 once more brought a candidate of the Colorado Party, Horacio Cartes, to power. The impeachment demonstrated the deep divisions within Paraguay and the unwillingness of elites to accept compromises for the benefit of the country's much poorer majority. Furthermore, the impeachment exposed how contested agribusiness, land rights and resource governance have become in Paraguay, but also showed the complex and contradictory regional relations evident in the reactions of South America's largest country Brazil and the regional organisations Unasur and Mercosur. These reactions are further evidence of the reluctance of governments across the region to address socio-environmental concerns related to resource governance at the regional level.

Significantly, the agribusiness lobby that has repeatedly blocked initiatives of land and tax reform in Paraguay and pushed for Lugo's impeachment is made up largely of cattle ranchers and soya exporters, many of whom are of Brazilian origin. Migration from Brazil to Paraguay has been promoted by both, Brazilian and Paraguayan governments since the 1960s when soya production in the region first started to expand. According to estimates, "brasiguayos" now produce over 85% of soybean in Paraguay. Brasiguayos have kept close ties with Brazil and benefitted from Brazilian credit and development aid as well as political support (Lambert 2016, 41–43). Together with wealthy Paraguayan farmers, brasiguayos constitute a powerful conservative political lobby that has not only contributed to significant environmental degradation in Paraguay, but has also on several occasions clashed with landless peasants (Lambert 2011a, b, 2012; Lambert and Nickson 2013; O'Shaughnessy and Ruiz Díaz 2009). Following the 2013 election, President Cartes lost no time in demonstrating his support for agribusiness and brasiguayo landowners (Lambert 2016, 44).

The Brazilian position vis-à-vis its small and historically isolated neighbour and its response to Lugo's impeachment reveals the inconsistent and ad hoc way in which South America's largest country has dealt with the transboundary dimensions of resource governance. On the one hand, disregarding the advice of the Brazilian Ministry of Foreign Affairs Itamaraty in 2009 President Lula agreed to the renegotiation of the terms of the Itaipú Treaty. The treaty had been signed by Brazil and Paraguay in 1973 when both countries were under military dictatorships. It set the conditions under which Paraguay could sell its excess electricity from the joint Itaipú hydropower station. The terms of the treaty were highly favourable to Brazil and had been regarded as unfair by Paraguayans for a long time.[6] Even if the renegotiated agreement was initially delayed in the Brazilian Senate and several key points were not addressed, it was seen as a major victory in Paraguay. Moreover, it constituted an important source of funding for social programmes, particularly given the lack of income from taxation. Lula's support for renegotiating the treaty was therefore a crucial element in strengthening the position of President Lugo (Canese 2013; Lambert 2011a, b, 182, 2016).

Furthermore, under President Rousseff, Brazil, together with other Latin American countries, criticised the impeachment process against Lugo on the grounds of constitutionality and procedure. Ignoring the objections of brasiguayos, Rousseff advocated the suspension of Paraguay from Unasur and Mercosur. This decision could have been interpreted as support for Paraguayan democracy and civil society, had it not paved the way for Venezuela's immediate entry into Mercosur. This had long been approved by the other three Mercosur countries, but stalled due to opposition in the Paraguayan senate. In line with the political and economic priorities of the other three Mercosur member states, Venezuela joined the bloc very shortly after Paraguay's suspension leading to suspicions that this was nothing more than a pretext to find a way for Venezuela's entry. This was followed by an upsurge of nationalist sentiment in Paraguay fuelled by a widespread perception that Mercosur served as an instrument of Brazilian hegemony leaving Paraguay in a dependent position with no political influence (Lambert 2012, 2016; Llanos et al. 2012, 6).

The expansion of agribusiness in the Southern Cone thus clearly demonstrates the regional dimensions of resource exploitation and their impact on regional politics and cooperation. Yet, government responses to the socio-environmental challenges arising from neo-extractivism are fragmented and inconsistent. This is particularly evident in the case of Brazil. Although Lula sought to find a compromise with Paraguay on transboundary energy questions, this has been on an ad hoc and bilateral basis rather than through regional cooperation. It is significant that regional organisations have not been given the mandate to address regional socio-environmental concerns resulting from intensive resource exploitation. Similarly, the decision to suspend Paraguay from the regional organisations was taken by the executives, but regional organisations have not been involved in terms of addressing the underlying issues that led to the impeachment process in the first place—despite their transboundary and regional nature. On the contrary, it seems that regional environmental concerns are getting less and less attention in regional organisations. While Unasur includes institutions covering a number of policy areas, including energy, health and social development, unlike the older Mercosur, there are none that are dedicated to environmental concerns. Other studies have found that, although Unasur has been keen to create new approaches to foreign policy and regional cooperation, this has not extended to the issue of climate change which is addressed infrequently and inadequately (Edwards and Roberts 2015, 23). Again this mirrors developments in Mercosur which, despite some engagement with global environmental regimes, has never become a key driver or even a very visible actor in these processes. Alba, on the other hand, has taken very radical stances on climate change emphasising climate justice and the historic responsibilities of industrialised countries, but the regional organisation has also been criticised for internal contradictions because governments have continued to rely on carbon-intensive resource exploitation in order to achieve national and regional objectives (Edwards and Roberts 2015, 101–134).

Overall, it is thus clear that natural resource exploitation has become central to the new forms of regional cooperation that have emerged over the last 15 years, but it is also one of the most controversial aspects. In this context, potential drivers for regional environmental cooperation have frequently been sidelined within regional organisations and integration processes. This is crucial to understand why the regional organisations and integration processes promoted by governments have not become frameworks for robust forms of regional environmental cooperation. Regionalism in South America is a mosaic of different regional organisations and initiatives with mandates that sometime overlap and sometimes compete (Tussie 2009, 170). However, in relation to regional *environmental* cooperation, there has been a trend that as the importance of resource exploitation for national and regional projects increased, environmental concerns and particularly those relating to resource exploitation have been addressed less and less in regional organisations. Given the importance of natural resource governance for neoliberal and post-neoliberal regionalism, it is difficult for governments to advance on regional environmental cooperation without addressing concerns related to resource exploitation. As a result, despite the relatively high level of institutionalisation of environmental concerns within Mercosur, regional organisations have not become a central framework for environmental cooperation, let alone a driver for this. Instead, as the following two chapters demonstrate, regional environmental cooperation has developed largely separate from regional organisations and in the margins of regional political agendas driven by networks of civil society organisations and NGOs, researchers, lower-level government officials and external funders. Perhaps, it is fair to claim that regional environmental cooperation has developed in spite of rather than because of government-led regional integration processes.

Notes

1. Interview government official, Secretaría de Ambiente y Desarrollo Sustentable, Buenos Aires, 2011, author's translation.
2. Following internal restructuring, the GTZ was renamed *Deutsche Gesellschaft für Internationale Zusammenarbeit*, or GIZ in 2011. However, as this took place after the end of the project that is relevant for this chapter, I use the term GTZ in the book.
3. Interview government official, Ministerio de Industria y Comercio, Asuncion, 2011, author's translation.

4. Interview former government official, Centro Tecnológico para la Sustentabilidad, Buenos Aires, 2011, author's translation.
5. Interview government official, Secretaría de Ambiente y Desarrollo Sustentable, Buenos Aires, 2011, author's translation.
6. As outlined in the following chapter, the Itaipú dam is one of the largest dams worldwide. It has 20 turbines each rated at 700 MW and each country is entitled to half of the electricity produced. However, Paraguay only uses about 7% of the energy output and the treaty stipulates that Paraguay sells its remaining electricity to Brazil at a fixed cost price rather than selling it at open market price or to a third party (Lambert 2011b, 182).

References

Briceño Ruiz, José. 2012. New Left Governments, Civil Society and Constructing a Social Dimension in Mercosur. In *Civil Society and the State in Left-Led Latin America—Challenges and Limitations to Democratization*, ed. Barry Cannon and Peadar Kirby. London: Zed Books.

Burges, Sean W. 2005. Bounded by the Reality of Trade: Practical Limits to a South American Region. *Cambridge Review of International Affairs* 18 (3): 437–454.

Câmara dos Deputados Brazil. 2010a. Parliamentarians Want to Prioritize the Environment in Mercosur Decisions. http://www2.camara.leg.br/english/chamber-of-deputies-news-agency/noticias/2010/2010-jul/parliamentarians-want-to-prioritize-the.

Câmara dos Deputados Brazil. 2010b. Parlasul Recommends an Environmental Preservation Area at the Triple Border. http://www2.camara.gov.br/english/chamber-of-deputies-news-agency/noticias/2010/2010jun/parlasul-recommends-an-environmental-preservation.

Canese, Ricardo. 2013. Itaipú: A Historic Achievement That Will Need to Be Closely Monitored. In *The Paraguay Reader*, ed. Peter Lambert, and Andrew Nickson. Durham and London: Duke University Press.

Carciofi, Ricardo. 2012. Cooperation for the Provision of Regional Public Goods: The IIRSA Case. In *The Rise of Post-Hegemonic Regionalism—The Case of Latin America*, ed. Pía Riggirozzi, and Diana Tussie. Dordrecht: Springer.

Chidiak, Martina. 2012. Investment Rules and Sustainable Development: Preliminary Lessons from the Uruguayan Pulp Mills Case. In *Rethinking Foreign Investment for Sustainable Development Lessons from Latin America*, ed. Daniel Chudnovsky, and Kevin Gallagher. Cambridge: Cambridge University Press.

Chodor, Tom. 2015. *Neoliberal Hegemony and the Pink Tide in Latin America*. Houndmills and New York: Palgrave Macmillan.

Chodor, Tom, and Anthea McCarthy-Jones. 2013. Post-Liberal Regionalism in Latin America and the Influence of Hugo Chávez. *Journal of Iberian and Latin American Research* 19 (2): 211–223.

CLAES. 2013. Centro Latinoamericano de Ecología Social. http://ambiental. net/claes/.

De Geus, Alex. 2011. *Las Caras de IIRSA: ¿integración Regional O Interconexión Sudaméricana?* https://www.academia.edu/2219320/Las_caras_de_IIRSA_ integración_regional_o_interconexión_Sudaméricana.

Devia, Leila. 1998. La Política Ambiental En El Marco Del Tratado de Asunción. In *Mercosur Y Medio Ambiente*, 2a ed. Leila Devia. Buenos Aires: Ediciones Ciudad Argentina.

Edwards, Guy, and J. Timmons Roberts. 2015. *A Fragmented Continent—Latin America and the Global Politics of Climate Change*. Cambridge, MA: MIT Press.

Emerson, R.Guy. 2014. An Art of the Region: Towards a Politics of Regionness. *New Political Economy* 19 (4): 559–577.

European Commission. 2007. *Mercosur—Regional Strategy Paper 2007–2013*. http://eeas.europa.eu/mercosur/index_en.htm.

FARN. 2013. Fundación Ambiente Y Recursos Naturales. http://www.farn.org. ar/.

Fulquet, Gastón. 2010. *Acuerdos de Integración Regional & Ambiente – Acciones Cooperativas Para Un MERCOSUR Sustentable*. 51. Documento de Trabajo. Buenos Aires: FLACSO Argentina. http://rrii.flacso.org.ar//web/wp-content/uploads/2010/11/FLA_Doc512.pdf.

Gardini, Gian Luca. 2011. Unity and Diversity in Latin American Visions of Regional Integration. In *Latin American Foreign Policies Between Ideology and Pragmatism*, ed. Gian Luca Gardini, and Peter Lambert. New York: Palgrave Macmillan.

Garzón, Jorge, and Almut Schilling-Vacaflor. 2012. *Infrastrukturprojekte Zwischen Geopolitischen Interessen Und Lokalen Konflikten*. 10/2012. GIGA Focus Lateinamerika. Hamburg: German Institute of Global and Area Studies. http:// www.giga-hamburg.de/de/publikationen/giga-focus/lateinamerika.

Giraudo, Maria Eugenia. 2014. *The Political Economy of Commodity Regions: The Case of Soybean in South America*. FLACSO-ISA Joint International Conference in Buenos Aires, Argentina, 23–25 July 2014.

Grugel, Jean. 2004. New Regionalism and Modes of Governance—Comparing US and EU Strategies in Latin America. *European Journal of International Relations* 10 (4): 603–626.

Grugel, Jean. 2007. Democratization and Ideational Diffusion: Europe, Mercosur and Social Citizenship. *Journal of Common Market Studies* 45 (1): 43–68.

Grugel, Jean, and Wil Hout, eds. 1999. *Regionalism Across the North-South Divide*. London: Routledge.

Hochstetler, Kathryn. 2003. Fading Green? Environmental Politics in the Mercosur Free Trade Agreement. *Latin American Politics and Society* 45 (4): 1–32.

Hochstetler, Kathryn. 2005. Race to the Middle: Environmental Politics in the Mercosur Free Trade Agreement. In *Handbook of Global Environmental Politics*, ed. Peter Dauvergne. Cheltenham: Edward Elgar.

Hochstetler, Kathryn. 2011. Under Construction—Debating the Region in South America. In *Comparative Environmental Regionalism*, edited by Lorraine Elliott and Shaun Breslin. Oxon: Routledge.

Hochstetler, Kathryn. 2012. Climate Rights and Obligations for Emerging States: The Cases of Brazil and South Africa. *Social Research* 79 (4): 957–982.

Hochstetler, Kathryn. 2013. South-South Trade and the Environment: A Brazilian Case Study. *Global Environmental Politics* 13 (1): 30–48.

Hochstetler, Kathryn, and Eduardo Viola. 2012. Brazil and the Politics of Climate Change: Beyond the Global Commons. *Environmental Politics* 21(5): 753–771.

Hogenboom, Barbara. 2015. New Elites around South America's Strategic Resources. In *Environmental Politics in Latin America—Elite Dynamics, the Left Tide and Sustainable Development*, edited by Benedicte Bull and Mariel Aguilar-Støen. Oxon: Routledge.

Kaltenthaler, Karl, and Frank O. Mora. 2002. Explaining Latin American Economic Integration: The Case of Mercosur. *Review of International Political Economy* 9 (1): 72–97.

Laciar, Mirta Elizabeth. 2003. *Medio Ambiente Y Desarrollo Sustentable*. Buenos Aires: Editorial Ciudad Argentina.

Lambert, Peter. 2011a. Dancing between Superpowers: Ideology, Pragmatism, and Drift in Paraguayan Foreign Policy. In *Latin American Foreign Policies Between Ideology and Pragmatism*, ed. Gian Luca Gardini, and Peter Lambert. New York: Palgrave Macmillan.

Lambert, Peter. 2011b. Undermining the New Dawn: Opposition to Lugo in Paraguay. In *Right-Wing Politics in the New Latin America: Reaction and Revolt*, ed. Francisco Dominguez, Geraldine Lievesley, and Steve Ludlam. London: Zed Books.

Lambert. 2012. The Lightning Impeachment of Paraguay's President Lugo. *E-International Relations*. http://www.e-ir.info/2012/08/09/the-lightning-impeachment-of-paraguays-president-lugo/.

Lambert, Peter. 2016. The Myth of the Good Neighbour: Paraguay's Uneasy Relationship with Brazil. *Bulletin of Latin American Research* 35 (1): 34–48.

Lambert, Peter, and Andrew Nickson. 2013. Epilogue: The Impeachment of President Fernando Lugo. In *The Paraguay Reader*, ed. Peter Lambert, and Andrew Nickson. Durham: Duke University Press.

Lenz, Tobias. 2012. Spurred Emulation: The EU and Regional Integration in Mercosur and SADC. *West European Politics* 35 (1): 155–173.

Llanos, Mariana, Detlef Nolte, and Cordula Tibi Weber. 2012. *Paraguay: Staatsstreich Oder „Misstrauensvotum"*? 8/2012 GIGA Focus Lateinamerika. Hamburg: German Institute of Global and Area Studies. http://www.giga-hamburg.de/de/publikationen/giga-focus/lateinamerika.

Luchetti, Javier Fernando. 2015. Political Dialogue in South America: The Role of South American Nations Union. In *Limits to Regional Integration*, ed. Søren Dosenrode. Farnham: Ashgate.

Lugo Méndez, Fernando. 2013. Inaugural Presidential Speech. In *The Paraguay Reader*, ed. Peter Lambert and Andrew Nickson. Durham: Duke University Press.

Lewis, Tammy L. 2016. *Ecuador's Environmental Revolutions: Ecoimperialists, Ecodependents, and Ecoresisters*. Cambridge, Massachusetts: MIT Press.

Malamud, Andrés. 2005. Mercosur Turns 15: Between Rising Rhetoric and Declining Achievement. *Cambridge Review of International Affairs* 18 (3): 421–436.

Malamud, Andrés. 2012. Moving Regions: Brazil's Global Emergence and the Redefinition of Latin American Borders. In *The Rise of Post-Hegemonic Regionalism—The Case of Latin America*, ed. Pía Riggirozzi and Diana Tussie. Dordrecht: Springer.

Malamud, Andrés. 2015. Interdependence, Leadership and Institutionalization: The Triple Deficit and Fading Prospects of Mercosur. In *Limits to Regional Integration*, ed. Søren Dosenrode. Farnham: Ashgate.

Malamud, Andrés, and Clarissa Dri. 2013. Spillover Effects and Supranational Parliaments: The Case of Mercosur. *Journal of Iberian and Latin American Research* 19 (2): 224–238.

Mercosur. 2003. Declaración de Principios de Producción Limpia Reunión de Ministros de Medio Ambiente de MERCOSUR. http://www.ambiente.gob.ar/?idarticulo-1145.

Mercosur. 2004a. *Elementos de Política Y Herramientas de Gestión Ambiental Y Producción Más Limpia En El Mercosur*. http://www.ambiente.gov.ar/?idarticulo=992.

Mercosur. 2004b. *Gestión Ambiental Y Producción Más Limpia En El Mercosur – Logros Alcanzados En Los Dos Primeros Años Del Proyecto Competitividad Y Medio Ambiente*. http://www.ambiente.gov.ar/?idarticulo=1669.

Mercosur. 2007a. *Política de Promoción Y Cooperación En Producción Y Consumo Sostenibles En El Mercosur*. Mercosur/CMC/DEC. No 26/07. http://www.ambiente.gob.ar/default.asp?IdArticulo=5480.

Mercosur. 2007b. *Gestión Ambiental Y Producción Más Limpia En El MERCOSUR – Logros Del Proyecto Competitividad Y Medio Ambiente*.

Mercosur. 2008. *La Temática Ambiental En El Mercosur: Evolución Y Perspectivas*. Rio de Janeiro.

Mercosur. 2012. *Declaración de Buenos Aires.* Acta No 01/12, Anexo III. http://www.ambiente.gov.ar/archivos/web/MERCOSUR/file/XVREUNIONDEMINISTROS/RMMA_2012_ACTA01_ANE03_ES_DeclaracionBuenosAires.pdf.

Moreno, Alicia. 2011. La Necesidad de Una Estrategia Ambiental En El MERCOSUR. *Densidades* (6): 63–77.

Monge, Jorge, and Klaus-Peter Jacoby. 2007. *Relevamiento Y Sistematización de Resultados E Impactos Del Proyecto "Competitividad Y Medio Ambiente" CyMA – MERCOSUR – Informe de Consultaría – Etapa Final.*

Newell, Peter. 2008. Trade and Biotechnology in Latin America: Democratization, Contestation and the Politics of Mobilization. *Journal of Agrarian Change* 8 (2 and 3): 345–376.

O'Shaughnessy, Hugh, and Edgar Venerando Ruiz Díaz. 2009. *The Priest of Paraguay—Fernando Lugo and the Making of a Nation.* London: Zed Books.

Phillips, Tom, and Mattia Cabitza. 2011. Bolivian President Evo Morales Suspends Amazon Road Project. *The Guardian,* September 27. http://www.guardian.co.uk/world/2011/sep/27/bolivian-president-suspends-amazon-road.

Riggirozzi, Pía. 2012a. Reconstructing Regionalism: What Does Development Have To Do With It? In *The Rise of Post-Hegemonic Regionalism—The Case of Latin America,* ed. Pía Riggirozzi and Diana Tussie. Dordrecht: Springer.

Riggirozzi, Pía. 2012b. Region, Regionness and Regionalism in Latin America: Towards a New Synthesis. *New Political Economy* 17 (4): 421–443.

Riggirozzi, Pía, and Diana Tussie. 2012. The Rise of Post-Hegemonic Regionalism in Latin America. In *The Rise of Post-Hegemonic Regionalism—The Case of Latin America,* ed. Pía Riggirozzi and Diana Tussie. Dordrecht: Springer.

Saguier, Marcelo. 2012a. Socio-Environmental Regionalism in South America: Tensions in New Development Models. In *The Rise of Post-Hegemonic Regionalism—The Case of Latin America,* ed. Pía Riggirozzi, and Diana Tussie. Dordrecht: Springer.

Saguier, Marcelo. 2012b. La Integración Sudamericana Y Los Conflictos Socio-Ambientales. FLACSO Argentina. http://www.flacso.org.ar/actividad_vermas.php?id=1591.

Scholvin, Sören, and Andrés Malamud. 2014. *Brasilien Als Geoökonomischer Knoten Südamerikas? 10/2014.* GIGA Focus Lateinamerika. Hamburg: German Institute of Global and Area Studies. https://www.giga-hamburg.de/de/publikationen/giga-focus/lateinamerika

Secretaría Administrativa del MERCOSUR. 2002. *Medio Ambiente En El Mercosur.* Montevideo. http://www.mercosur.int/show?contentid=464&channel=secretaria.

Serbin, Andrés. 2012. New Regionalism and Civil Society: Bridging the Democratic Gap? In *The Rise of Post-Hegemonic Regionalism—The Case of Latin America,* ed. Pía Riggirozzi, and Diana Tussie. Dordrecht: Springer.

Somos Mercosur. 2013. Somos Mercosur. http://www.somosmercosur.net/.

Telò, Mario. 2006. *Europe: A Civilian Power?*. Houndmills: Palgrave Macmillan.

Torres, Alicia, and José Pedro Diaz. 2011. MERCOSUR Ambiental: ¿se Trata de Una Mirada Sólo Desde El Comercio O Del Avance de La Dimensión Olvidada? ¿Medio Lleno O Medio Vacío? In *Mercosur 20 Años*, ed. Gerardo Caetano. Montevideo: CEFIR. http://cefir.org.uy/documentacion/publicaciones-cefir.

Turzi, Mariano. 2011. The Soybean Republic. *Yale Journal of International Affairs* (Spring-Sum): 59–68.

Tussie, Diana. 2009. Latin America: Contrasting Motivations for Regional Projects Latin America. *Review of International Studies* 35: 169–188.

Tussie, Diana, and Patricia Vásquez. 2000. Regional Integration and Building Blocks: The Case of Mercosur. In *The Environment and International Trade Negotiations—Developing Country Stakes*, ed. Diana Tussie. Houndmills: Macmillan Press Ltd.

UNEP, and CLAES. 2008. *GEO Mercosur – Integración, Comercio Y Ambiente*. Montevideo.

Villar, Pilar Carolina, and Wagner Costa Ribeiro. 2011. The Agreement on the Guarani Aquifer: A New Paradigm for Transboundary Groundwater Management? *Water International* 36 (5): 646–660.

Wachendorfer, Achim. 2013. *Wieder Alles Beim Alten? Paraguay Nach Den Wahlen*. Friedrich Ebert Stiftung. http://library.fes.de/pdf-files/iez/10006.pdf.

CHAPTER 4

Regional Environmental Cooperation in the La Plata River Basin

Although the regional organisations which have been the focus of most research on regional cooperation in South America have not become a driver for regional environmental cooperation in the Southern Cone, robust forms of environmental cooperation have developed in another regional framework, notably the La Plata basin regime, an example of a regional resource regime. This type of regional regime has received much less attention and has generally not been considered much in the literature on regional integration. The La Plata basin regime consists of a network of treaties and technical commissions that were created in the late 1960s to early 1970s in order to promote peace, security and economic development in the five riparian countries Argentina, Bolivia, Brazil, Paraguay and Uruguay. Following changes in the global and domestic context, it became a framework for regional environmental cooperation two decades later when a series of treaties were signed and a number of projects were developed focussing specifically on ecological aspects of the basin. This was promoted by a combination of endogenous and exogenous drivers who fostered joint research, monitoring and environmental protection activities in the basin. In several cases, the work of regional university networks was important in laying the basis for environmental cooperation by developing a better understanding of environmental problems in the basin and the transboundary nature of these. In addition, international funders, most importantly the GEF, have been crucial in supporting several environmental projects in the basin. Over the last 25 years, cooperation on shared environmental concerns in the La Plata basin has thus increased, both in formal agreements and

© The Author(s) 2017
K.M. Siegel, *Regional Environmental Cooperation in South America*,
International Political Economy Series, DOI 10.1057/978-1-137-55874-9_4

activities in practice. Nevertheless, important limitations remain, demonstrating the marginality of environmental cooperation. In particular, there is a high dependence on external funding and environmental concerns often clash with other objectives, notably economic development. In addition, environmental regulations are often vague and difficult to implement. In addition, although various civil society networks have demonstrated an interest in the governance of the basin, they have been sidelined in decision-making processes and in the projects carried out. As a consequence regional civil society networks have been more influential in protesting against planned large-scale projects in the basin resulting in delays, downsizing and reversing of such plans than in driving environmental cooperation in the basin.

THE LA PLATA BASIN REGIME

The La Plata basin is one of the five biggest basins worldwide and the second biggest in South America after the Amazon. It consists of a number of sub-basins which flow into the Rio de la Plata and eventually into the Atlantic Ocean. Overall, the La Plata basin covers a big part of central and northern Argentina, southeast Bolivia, almost the whole of southern Brazil, all of Paraguay and a big proportion of Uruguay. In addition to the surface waters, ground waters are an important part of the basin, among them the Guaraní aquifer, one of the largest underground water reserves worldwide which is shared between all the La Plata basin countries except Bolivia (del Castillo Laborde 2008; Elhance 1999, 25–52; Gilman et al. 2008; Tucci and Clarke 1998).

The basin is at the economic and political centre of each of the five basin countries. The capital cities of all five riparian countries are located either on the banks of one of the basin's rivers or in the catchment area of the basin. As a consequence, about 70% of the per capita gross domestic product of the five countries is generated in the area of the basin. Moreover, with a number of dams, including some of the largest dams worldwide, the rivers are crucial for the generation of energy as well as transportation (Pochat 2011, 497–498; Tucci and Clarke 1998). Modifications of the rivers for these purposes as well as urban, industrial and agricultural pollution are among the main environmental concerns, and, in many cases, they directly impact on the quality of life of the people living in the basin. In 2007, the WWF listed the La Plata as one of the ten most threatened rivers worldwide (Gilman et al. 2008, 208; Tucci and Clarke 1998; WWF 2007). However, while there has been significant international pressure to preserve the Amazon basin, the La Plata basin has received comparatively little attention and the politics of environmental cooperation in the La Plata basin have been researched in much less detail.

Cooperation in relation to the La Plata basin accelerated significantly in the late 1960s and 1970s. An overall framework was provided by the La Plata Basin Treaty signed in 1969 between the five countries sharing the basin. The overall objective of the treaty is the "balanced and harmonious development and the physical integration" of the basin in particular in relation to navigation; the rational use of water; the preservation of animal and plant life; transport, energy and communication infrastructure; promotion of industry; education and health; knowledge and other projects of common interest (Tratado de la Cuenca del Plata 1969, author's translation). In addition, the Intergovernmental Coordinating Committee of the La Plata Basin Countries (Comité Intergubernamental Coordinador de los Países de la Cuenca del Plata or CIC) was set up with a permanent secretariat in Buenos Aires as a mechanism to implement the treaty. With political and technical representatives from all five countries, the CIC is an overarching multilateral commission for the basin as a whole. It is important for environmental cooperation in the basin as one of its main tasks has been to coordinate the different initiatives of environmental cooperation in the basin and centralise information on these. Moreover, in 1974 the Development Fund of the La Plata Basin (Fondo Financiero para el Desarrollo de la Cuenca del Plata or FONPLATA) was created as a funding mechanism. In addition to these basin-wide instruments, several other bilateral or trilateral treaties were signed for specific areas of the basin and in many cases complemented with technical commissions to implement the treaty objectives (del Castillo Laborde 2008, 277–278; Gilman et al. 2008, 205; Pochat 2011, 500). Although there are some variations in the organisational structures and tasks, the river commissions generally consist of technical experts designated by the riparian countries. This web of different treaties and technical commissions makes up the La Plata basin regime.

The La Plata basin regime is an example of a regional resource regime where interdependency and the desire to develop joint projects of mutual benefit were the main motivations behind the creation of the regime. During the 1960s and 1970s, the focus of cooperation was mainly on economic development and in particular the generation of energy. For governments in the region, a key objective at the time was to achieve the level of economic development of the industrialised countries (Kempkey et al. 2009, 262). Moreover, following the 1973 oil shocks, the countries were looking for independent sources of energy (Da Rosa 1983, 79; Elhance 1999, 31; Gilman et al. 2008, 207). To achieve these objectives, cooperation between the basin states was necessary for two main reasons. First, the projects that were planned were enormous and several of them were in border areas and therefore required the cooperation of at least two of the basin countries. With a dam 196 m high and just under eight kilometres long and a reservoir covering an area of 1.350 km^2, the Itaipú dam between Brazil and Paraguay was for a long time the world's largest power station, but it was overtaken in terms of installed capacity in 2012 when the Three Gorges power station in China became fully operational (Elhance 1999, 25–26; Gilman et al. 2008, 207; Itaipu Binacional 2016).[1] This means one of the reasons for the development of the La Plata basin regime was the prospect of developing mega-projects with benefits for two or more countries although the distribution of benefits and burdens of these projects between and within countries remains a contested question as discussed below.

Second, the regime established in the late 1960s and 1970s also played a crucial role in ensuring stability by providing written agreements to regulate the usage of the rivers. This is important because projects developed in one part of the basin can have major impacts on downstream countries. The period of the 1960s and 1970s was characterised by military dictatorships and a climate of mutual distrust, and this was also evident in relation to the usage of the rivers.[2] The rivers of the La Plata basin thus did not only provide a valuable resource, but also led to "bitter confrontation between riparian states" (del Castillo Laborde 1999, 183), in particular between the two biggest and most powerful states of the basin, Argentina and Brazil, which had been competing for regional dominance since gaining independence. Cooperation in this phase was very successful in the sense that it achieved its aims of promoting economic development and stability (Gilman et al. 2008, 207–208). The La Plata Treaty also opened up new communication channels and set a precedent for cooperation in the region in a broader sense by providing a channel to resolve conflicts politically rather than militarily (Kempkey et al. 2009, 269). Overall, the La Plata basin is a central feature defining the Southern Cone region and its waterways and hydropower stations play an important role in the physical integration of the region.

Environmental Cooperation in the La Plata Basin

Cooperation on environmental concerns was not a priority when the La Plata basin regime was first created, but with a changing global and domestic context, it gained in importance from the 1990s onwards. At the global level, there was a shift in the 1990s towards taking environmental concerns more into account in water governance while still pursuing economic development. This was reflected also in the principle of "sustainable development" which became a central paradigm for environmental governance in that time and which promised to provide a way to address both, economic development and environmental sustainability. This framing made environmental protection compatible with the objectives of the Southern Cone governments whose focus at that time was on economic growth as discussed in Chap. 2. Domestically, the return to democracy in the previous decade increased the possibilities for civil society activity. From the 1990s onwards, there are several examples of civil society protests which put pressure on governments to take socio-environmental concerns into account in the governance of the La Plata basin. As a result of protests by affected communities and civil society networks, several large infrastructure projects in the basin were stopped, changed or delayed. In this context of a greater openness towards taking environmental concerns into account on the part of governments and civil society pressure to do this, a series of new agreements where environmental concerns have gained a more prominent position was signed by governments. In addition, a series of projects were initiated, for the large part funded by the GEF, but developed and implemented in cooperation with regional research networks.

To some extent, the development of a more favourable climate for environmental cooperation in the La Plata basin reflected global trends. In most river basins, social and environmental costs were generally not considered much in the first projects, and these focussed mostly on issues such as transportation, hydropower generation, irrigation or flood control. However, this started to change from the 1970s onwards. With the rise of the concept of integrated water resource management in the 1990s, ecological and social concerns moved to the centre of attention (Molle 2009). During the same time period, the principle of sustainable development developed into a key guiding norm for environmental governance. This was crucial for countries of the South because the concept seemed to offer a

way of reconciling the objective of economic development with environmental concerns (Vogler 2007).

Similar to other parts of the world, initially environmental concerns were
barely taken into account in the La Plata basin regime. Consequently, the
success of the regime in terms of promoting stability and economic development was dependent on significant environmental costs (Gilman et al.
2008). In addition, there were significant negative effects of previous
developments in the basin for people living in affected areas. It is not a
surprise that when the return to democracy increased the possibilities for
citizens to express their concerns, there were multiple protests against new
large-scale projects in different parts of the basin. Despite the abundance of
freshwater in the region and the many projects that had already been
developed in the basin, it has been estimated that in the mid-1990s around
20% of the population in several of the basin countries did not have access
to safe drinking water, although there were important geographic variations (Elhance 1999, 32). Moreover, the social and environmental costs
and benefits of large-scale projects that were carried out in the La Plata
basin previously were often distributed highly unequally with those groups
with less political and social power, such as indigenous groups and the
poor, bearing a disproportionately large share of the disadvantages. Large
hydropower installations, for example, required the flooding of large areas
with severe impacts on the ecosystem and leading to the displacement of
tens of thousands of people (Bartolome and Danklmaier 2012; Elhance
1999, 36). Resentment of such projects was made worse by the fact that in
particular for projects developed under military dictatorships participation
of affected communities in planning was non-existent. Moreover, large-
scale projects like the Yacyretá dam between Argentina and Paraguay and
the Itaipú dam between Brazil and Paraguay acquired a reputation for
mismanagement, over-inflated costs and corruption which meant they
benefited elites and companies engaged in their construction disproportionately while having significant negative impacts for people living in those
areas (Elhance 1999, 40–47; O'Shaughnessy and Ruiz Díaz 2009, 86–87).
In addition, the extent of the costs for individuals can be very uncertain.
Studies in Argentina, for example, have found that there is no general
policy framework regulating the displacement and resettlement of communities affected by large-scale development projects, so that each project
develops its own policy for resettlement and compensation and in the past
this has been rather chaotic in some cases (Bartolome and Danklmaier
2012, 124). Overall, this means that there was already a history of negative

experiences and a number of reasons for dissatisfaction and suspicion of large-scale projects prior to the return to democracy. The following three examples demonstrate how affected or potentially affected communities together with environmental NGOs put pressure on decision-makers, and as a result, several large-scale projects that were planned in the La Plata basin in the 1990s and 2000s were stopped, downscaled or delayed.

A first example was the strong opposition to the so-called hidrovía, or water superhighway, which was planned during the late 1980s and aimed to make over 3000 km of the La Plata basin navigable all year round. This would have had significant social costs and a large environmental impact in particular on the Pantanal, one of the world's largest wetlands. Because of this, an alliance of local and international NGOs came together in the 1990s opposing the project. This alliance came from a wide range of backgrounds including environmentalists, indigenous groups, unions and professional organisations as well as universities and research centres. The coalition emphasised the negative social and environmental impact of the project which presented a threat to jobs and ways of life as well as ecosystems, biodiversity and water quality. Following the protests, the IDB eventually withdrew its funding for the project which stopped the initial plan, although individual parts of the hidrovía have been developed (Bucher and Huszar 1995; Elhance 1999, 48–49; Gottgens et al. 2001; Hochstetler 2002; Tussie and Vásquez 2000, 194).

Public opinion and opposition of citizens have also prevented the construction of another major bi-national hydropower dam on the Paraná River between Argentina and Paraguay. The Corpus Christi dam had been planned since the 1970s, but its construction was delayed for various reasons. It became a subject of debate in local politics and was put to a referendum in the Argentinean province of Misiones in 1996. The vote resulted in an overwhelming majority against the dam. In this case, the legacy of the past and the unequal distribution of costs appeared to have had a large impact. Technical staff maintain that the plans for the Corpus Christi project have a much smaller environmental and social impact than earlier hydropower dams and mitigation strategies have been studied extensively in the initial assessments. However, because previous projects and in particular the Yacyretá dam on the same river had such devastating social and environmental consequences, people fear this might happen again.[3] According to one technical expert:

When Yacyretá was decided, it was mostly a geopolitical topic...And in those years, in the 1970s, the topic of the environment was not the same as it is now. So really from an environmental point of view it was a disaster...And there they flooded thousands of hectares. And they left the people without a place to stay, it brought thousands of problems. That was the big problem. The people from Misiones who suffered, they don't want another Yacyretá. The thing is...with Corpus this wouldn't happen. First of all because the impact is much smaller. And second because everything is considered before. Not as it was done with Yacyretá. For Yacyretá they built the dam and then they thought about the environment. Here it's different, from the beginning the environment comes first and then comes the topic of the construction.[4]

In many respects, the debate over the Corpus dam reflects the highly contested nature and the lack of consensus with regard to hydropower projects in the basin. Civil society members have raised not only the question of environmental damage, but also the complex question of the distribution of benefits from large hydropower projects. According to one interviewee, electricity should be free in Paraguay due to the large capacity of the Itaipú dam,[5] and another interviewee noted that public transport in Paraguay should be electric.[6] Yet, most of the energy produced in Paraguay is exported to economic centres in neighbouring Argentina and Brazil according to conditions that were initially set during the military dictatorships and that are still regarded as highly unfair by many Paraguayans although the renegotiation of the Itaipú treaty discussed in Chap. 2 has improved this a bit. As governments look for new sources of energy in order to satisfy growing domestic demands and support economic growth, the development of hydropower projects continues to be a controversial issue which is closely related to larger questions of what development and regional integration should look like, who they should benefit and how decisions should be taken.[7]

Similar questions were at the heart of the large-scale and highly visible protests that developed in Argentina in the 2000s against the construction of a large pulp mill by the Finnish company Botnia (now UPM). The mill was built on the Uruguayan side of the Uruguay River which in this section forms the border between Argentina and Uruguay. The protests on the Argentinean side led to an international conflict between the two countries. Although the protests were not successful in preventing the construction and operation of the targeted mill, they were nevertheless influential in other respects. Most importantly, initially the Uruguayan government had granted permission for the construction of two pulp mills. Yet, following the controversies over the Botnia mill, the Spanish company ENCE changed its plans, so that only one pulp mill was constructed in the original location (Bueno 2010, 171–187; Waisbord and Peruzzotti 2009). There is also some evidence that, fearing public disapproval, other smaller pulp mills operating in the region started to upgrade their production processes (Dudek 2013, 118–120). Finally, the high visibility and the large political impact of the protest helped raise public awareness of environmental concerns in the La Plata basin more generally.

Overall, this means that protests by affected communities and environmental groups benefiting from the greater openness that democratisation brought have resulted in greater pressure on decision-makers to take socio-environmental concerns into account. With few effective channels for participation in decision-making, civil society groups have often resorted to "outside strategies" (Uhlin 2011, 854) such as protests, or applied pressure by using the "boomerang strategy" and cooperating with Northern NGOs (Hochstetler 2002). The main impact of this has been in stopping, delaying and/or downscaling planned developments in the basin and generally raising awareness of the socio-environmental dimensions of resource governance in the La Plata basin. Overall, from the 1990s onwards, pressure on governments to address the socio-environmental concerns of the basin thus increased at the international and the domestic level. In parallel, regional research networks that had been working on ecological aspects and international funders both promoted transboundary cooperation on environmental concerns.

In relation to formal cooperation, a shift occurred in the 1990s when several more treaties were signed (see Table 4.1 for an overview of all agreements since 1969). This time environmental sustainability became one of the key aspects, and some agreements refer to specific environmental concerns such as soil conservation or water quality (del Castillo Laborde 2008, 281–282; Gilman et al. 2008, 208; Kempkey et al. 2009, 271; Pochat 2011, 505).

This was complemented with a significant increase in cooperation in practice in the same time period. From the 1990s onwards, six large environmental projects with external funding have thus been carried out in different parts of the basin.[8] With the exception of the Pantanal and Upper Paraguay River Basin project, which was mostly carried out in Brazil and only had smaller international components (OAS 2005a), all of these projects have involved government agencies in at least two countries, thus

Table 4.1   Overview of the most important agreements in the La Plata basin 1946–2010

Year	Agreement	Countries	Main points
1969	La Plata Basin Treaty	Argentina, Bolivia, Brazil, Paraguay, Uruguay	Establishment of a framework for a balanced and harmonious multilateral development and utilisation of the basin's water resources
1971	Agreement for the study of the development of the Parana River resources	Argentina, Paraguay	Creation of the Argentinean-Paraguayan Joint Commission of the Parana River (COMIP) for the administration of the shared stretch of the river and the development of the Corpus Christi multiple-purpose hydraulic project
1973	Itaipú Treaty	Brazil, Paraguay	Creation of Itaipú Binational for constructing Itaipú hydropower development
1973	Treaty on the La Plata River and its Maritime Front	Argentina, Uruguay	Settlement of a controversial situation about the exercise of jurisdiction over the river's waters; the treaty also deals with navigation, fishing, bed and subsoil, pollution prevention and other issues and sets up the Administrative Commission for the La Plata River (CARP) and the Joint Technical Commission for the Maritime Front (CTMFM)
1973	Yacyretá Treaty	Argentina, Paraguay	Creation of Yacyretá Binational Entity (EBY) for constructing Yacyretá hydropower development
1974	FONPLATA Constituting Agreement	Argentina, Bolivia, Brazil, Paraguay, Uruguay	Creation of the Financial Fund for the Development of the La Plata Basin (FONPLATA) in order to lend financial support to the activities envisioned in the La Plata Basin Treaty

(continued)

Table 4.1 (continued)

Year	Agreement	Countries	Main points
1975	Uruguay River Statute	Argentina, Uruguay	Creation of the Administrative Commission for the Uruguay River (CARU) for dealing with the regulation of navigation, works, pilotage, bed and subsoil resources, fishing, pollution prevention, jurisdiction and settlement of dispute procedures
1979	Tripartite Agreement on Corpus and Itaipú	Argentina, Brazil, Paraguay	Setting of the maximum operating level for Corpus Christi dam and conditions for the operation of Itaipú power plant
1980	Binational Boundary Treaty	Argentina, Brazil	Agreement on the use of their shared stretch of the river; establishment of principles related to energy production, mitigation of extraordinary floods, improvement of navigation, water uses, and keeping of health conditions
1989	Resolution of the Foreign Affairs Ministers incorporating the Paraguay-Parana Waterway Programme to the La Plata Treaty System	Argentina, Bolivia, Brazil, Paraguay, Uruguay	Creation of the Intergovernmental Committee for the Paraguay-Parana Waterway, Cáceres Port-Nueva Palmira Port (CIH)
1991	Cooperation Agreement for the Utilisation of the Natural Resources and the Development of the Cuareim/Quaraí River Basin	Brazil, Uruguay	Creation of the Joint Uruguayan-Brazilian Commission for the development of the Cuareim/Quaraí River Basin (CRC)
1993	Pilcomayo Lower Basin Agreement	Argentina, Paraguay	Creation of the Administrative Binational Commission of the Lower Basin of the Pilcomayo River, for its integral management, including use and regulation of discharges,

(continued)

Table 4.1 (continued)

Year	Agreement	Countries	Main points
			project and execution of works and water quality
1995	Pilcomayo Trinational Commission Constituting Agreement	Argentina, Bolivia, Paraguay	Creation of the Trinational Commission for the development of the Pilcomayo River basin
1995	Agreement for the Multiple Development of the Resources of the Upper Basin of the Bermejo River and the Grande de Tarija River	Argentina, Bolivia	Creation of the Binational Commission for the Development of the Upper Basin of the Bermejo River and the Grande de Tarija River (COBINABE)
2006	Cooperation Agreement for the Sustainable Development and Integrated Management of the Apa River Hydrographic Basin	Brazil, Paraguay	Creation of the Brazil-Paraguay Commission of the Apa River Basin
2010	Guarani Aquifer Agreement	Argentina, Brazil, Paraguay, Uruguay	Reaffirmation of the sovereignty of the four states, inclusion of norms of general international law that regulate the use of shared natural resources and creation of a Commission with the aim of coordinating the cooperation

Most important agreements in the La Plata basin (1946–2010), adapted from Gilman et al. (2008) and Pochat (2011)

making a clear contribution to regional environmental cooperation in the La Plata basin. While most projects have addressed surface waters, that is the different rivers of the La Plata basin, the Guaraní aquifer project related to underground water resources shared by four countries (OAS 2005b). The GEF is the most important funder for these projects and has provided funding for all projects, with the exception of the Pilcomayo River which received support from the EU. Several of the larger projects were preceded by the work of regional research networks who were also involved in the first project proposals involving partners in different countries.

The various projects have been carried out in different parts of the basin with large variations in both the environmental problems to be addressed and the socio-economic context. Nevertheless, the areas of cooperation in the different projects are relatively similar. In all projects, an important component is thus the shared generation and management of information through joint research and monitoring. The FREPLATA project between Argentina and Uruguay in relation to a part of the Rio de la Plata included research projects with universities for pollution prevention and control as well as initiatives to raise awareness and promote public participation (del Castillo Laborde 2008, 285). In the north of the basin, a project for the Pilcomayo River shared by Bolivia, Paraguay and Argentina laid the basis for joint monitoring and a flood warning system which still continue to operate after the end of the project. In addition, information about the activities of the Trinational Commission for the Pilcomayo River is publicly accessible on its website which is regularly updated. This includes technical information such as data and readings from the different monitoring stations, meeting documents and an online library although access to some of this information requires authorisation through the Trinational Commission.[9]

In the case of the Guaraní aquifer, the existence of underground water resources had been known for some time, but it was only in the 1990s that the transboundary nature of the aquifer was discovered and researchers realised that what they thought were separate aquifers, was in fact one connected aquifer. This initiative initially received funding from a Canadian agency, but when this finished the group of universities that had been working on the topic proposed a regional research project to increase the knowledge of the aquifer and presented this to the World Bank.[10] The World Bank decided that it was not its role to finance a research project. At the same time, however, the governments of Brazil and Uruguay through the Organization of American States (OAS) had presented a proposal to the GEF for a joint project for the management of the Cuareim River, a tributary of the Uruguay River, shared between the two countries. The GEF considered that the Cuareim River was very small and not of global importance. Although both these project proposals were turned down individually, because they were developed in the same period of time and in relation to related topics, the GEF analysed the proposals together and eventually this led to the GEF project on the Guaraní aquifer.[11] This clearly shows how two elements, the work of regional research networks and the GEF as a new funding mechanism for international waters, coincided and produced a window of opportunity that advanced regional environmental cooperation in the La Plata basin. The project on the Guaraní aquifer was very successful in significantly increasing the scientific knowledge of the aquifer system (Giraut et al. 2010, 3; Sindico 2011, 257; Villar and Ribeiro 2011, 649). Moreover, in at least one case, a pilot project established in the framework of the GEF project has resulted in continuous cooperation lasting beyond the GEF project. Led by a group of people committed to maintaining regular exchanges of information and dialogue between the city of Salto in Uruguay and the neighbouring city of Concordia in Argentina, cooperation in practice has thus continued after the end of the GEF project albeit at a smaller scale (Sindico 2016).

Finally, the so-called framework project, which relates to the basin as a whole and is carried out by the CIC, aims at monitoring and controlling the effects of climate variability such as floods and droughts, both of which have caused severe problems in parts of the basin. A second important objective of the project is to centralise information from all the projects in the basin (CIC 2009, 3; del Castillo Laborde 2008, 284; OAS 2005c; Pochat 2011, 502, 505). On the whole, the framework project is therefore important for promoting a dialogue on environmental concerns in the La Plata basin between technical experts and political representatives from the five countries. Moreover, as one interviewee pointed out, the project is also important because it strengthens the CIC itself:

> That's the objective that the countries have today, that this represents a strengthening of the role that the Treaty of the La Plata basin has given to the CIC. Because the treaty says that the CIC has to coordinate what is called the system of the La Plata basin.[12]

The idea for the framework project partly built on research on climate-related aspects of the basin that public universities in the region had been working on since the 1990s.[13] The project became concrete at the start of the new millennium with the CIC acting as the local institution executing the project.

Generating and sharing information through joint research and monitoring are therefore crucial elements of cooperation in practice in the La Plata basin. Some of the projects have also included smaller pilot projects on specific issues, and many of them have included components for awareness raising and environmental education. Moreover, in most cases, the projects were only seen as a first step in order to develop a more comprehensive water management strategy to be implemented in the long run. The overall objective of the Guaraní aquifer project, for example, was the development and implementation of an institutional, legal and technical framework to preserve and manage the Guaraní aquifer system (OAS 2005b). By the end of the project in 2009, this had not yet been achieved, but a year later the heads of state of the four countries used the occasion of a Mercosur meeting to sign the agreement on the Guaraní aquifer (Villar and Ribeiro 2011, 651) although five years later this had still only been ratified by Argentina and Uruguay (Villar 2016, 16). In relation to the Pilcomayo River, the main outcome of the cooperation project with the EU was a master plan for the integrated management of the basin. The aim of the plan is to strengthen the process of transboundary integration by addressing concerns shared by the three countries such as water quality and availability of water, erosion and sedimentation, fish, risk management, establishment of monitoring systems, exchange of experience, communication and dissemination of information, institutional strengthening and sustainable economic and human development (Proyecto de Gestión Integrada y Plan Maestro de la Cuenca del Río Pilcomayo 2008, III, 11).

Similarly, the first stage of the framework project was to develop an analysis bringing together important information, outlining the main challenges and developing a common vision of how the basin should be managed with an emphasis on environmental sustainability. This then serves as a basis for developing a more detailed strategy. For the FREPLATA project too, the main outcome of the first phase was the development of an analysis which serves as the base document for the development of a more concrete action programme to prevent and mitigate transboundary environmental problems in the area (FREPLATA 2005). The objective of the second phase was to reduce and prevent pollution from land-based sources through the implementation of the action programme developed after the first phase. Although the issues addressed and the institutional frameworks are very different, in several respects, environmental cooperation in this case study has followed similar processes to those on migratory species examined in the following chapter. In both cases initially joint research and monitoring was important to understand the nature of the environmental problems and the transboundary elements in this better. As a second stage, the development of joint plans to address this follows. Moreover, in both cases, environmental cooperation focuses mostly on technical framings and solutions.

Overall, the different projects in the La Plata basin have resulted in some important achievements. The second phase of the FREPLATA project involves 37 key stakeholders, including nine ministries, navy, coast guards, provincial and local authorities, and private sector representatives. This is the broadest support of a strategic action programme in the history of the UN Development Programme (UNDP) International Waters programmes and also an important achievement for the GEF (GEF 2009, 9–10, 16–17). In addition, the project made an important political achievement in a very different way. While Argentina and Uruguay were engaged in the pulp mill conflict, the FREPLATA project was delayed, but ultimately continued. As one interviewee noted, the project therefore significantly contributed to maintaining a dialogue on the topic of environmental concerns in a shared river, even though the two countries were at the same time involved in a serious dispute over the very same topic:

> ...despite this problem, Argentina and Uruguay continue to work on the Rio de la Plata and its maritime front without any political problems. This project achieved this, that Uruguay and Argentina could continue to work despite the conflict on this same topic in another international body of water. The two governments signed the strategic action plan in the middle of the conflict on the Uruguay river.[14]

In relation to the Pilcomayo River, what stands out is the emphasis on public participation and the way this is institutionalised in the basin management. The institutional set-up thus includes a political and a technical entity, as well as a Trinational coordination committee (Comité de Coordinación Trinacional) which is made up of five representatives of each country to ensure the participation of civil society (Proyecto de Gestión Integrada y Plan Maestro de la Cuenca del Río Pilcomayo 2008, 11). According to staff of the Trinational coordination committee, the Pilcomayo management structure is exceptional not only in the La Plata basin regime, but also worldwide.[15]

Similarly, the treaty on the Guaraní aquifer is noteworthy because it is one of only a few agreements signed in relation to transboundary underground water resources worldwide (Cassuto and Sampaio 2011, 664). One interviewee noted that the Guaraní aquifer is now a model, not only for the region, but also globally.[16] The preamble of the Guaraní aquifer agreement refers to Resolution 63/124 of the UN General Assembly on the Law of Transboundary Aquifers and reflects the main principles and commitments of international law: equitable use of water resources, the obligation not to cause harm and cooperation (Sindico 2011, 266; Villar and Ribeiro 2011, 654). It is also worth noting that the agreement was not sparked by any transboundary conflicts over the aquifer and the initiatives taken are mostly precautionary, as so far both exhaustion and pollution of the aquifer are not a major problem yet, but they are concerns for the future (Cassuto and Sampaio 2011, 661–662; Sindico 2011, 257; Villar and Ribeiro 2011, 646).

Overall, it is evident that since the 1990s environmental concerns have been given more attention in both formal treaties and cooperation projects. Regional university networks and donors have been crucial in terms of promoting specific initiatives of regional environmental cooperation in the basin, and these have led to some noteworthy achievements. Nevertheless, important limitations and challenges also remain.

LIMITATIONS, CONTESTATIONS AND SIGNS OF MARGINALITY

Despite the important achievements noted above, environmental cooperation in the La Plata basin remains marginal. Regulations and agreements are often contradictory or vague making them hard to implement, and many of the projects noted above have experienced significant delays. Moreover, whereas cooperation on other aspects, notably improving the transport and energy infrastructure of the basin, has received funding from sources within the region, cooperation on environmental concerns is still highly dependent on external funding and the GEF in particular. As a consequence, external funders have been quite influential in determining the nature of cooperation in the basin. At the same time, regional civil society networks which have expressed a clear interest in the governance of the basin and which constitute important potential endogenous drivers for environmental cooperation have been sidelined. Although most projects include some components for civil society engagement, this has not translated into notable influence in decision-making. In addition, governments decided to give the mandate to deal with the basin governance to the less accessible and more technical institutions of the La Plata basin regime rather than the regional organisation Mercosur which initially had the potential to be more accessible. This has further limited possibilities for civil society participation.

In most projects, the first step of environmental cooperation has consisted of an analysis of the existing environmental problems and the institutions to address these. In several cases, this initial analysis uncovered unresolved conflicts between different objectives, most importantly economic aims and environmental concerns. What is more, the division of responsibilities between different authorities at the national and the regional level is often unclear or very complex. This means moving to the next step in environmental cooperation and creating more effective institutions is much more difficult.

As an interviewee involved in the FREPLATA project explained, the second phase of the project is much more political. Whereas the first phase brought together important technical information, the aim of the second phase is now to define how to use this information and set up the necessary institutional structures in both countries. One of the main challenges of the second phase is to achieve that the governments take over the work and responsibilities in the long term rather than just relying on the project.[17] One of the political challenges that the initial analysis identified is the lack of coordination between different regulations. Many instruments overlap, but on the other hand, the objectives are sometimes contradictory. Moreover, the analysis points out that sometimes the implementation of regulations is, in fact, impossible because the appropriate technology is not available and a strict application would lead to the closure of major industrial plants with high social costs. Consequently, many regulations are not implemented at all. Moreover, as a result of the state reforms of the 1990s, staff was cut down leading to a reduced capacity for monitoring and control. The results of this incoherent system of regulations and instruments are that the general public and the private sector do not receive clear signals nor adequate incentives for better environmental management (FREPLATA 2005, 203–210).

The agreement on the Guaraní aquifer equally has been criticised for the lack of specific arrangements with regard to protection and water extraction as well as enforcement. Although the treaty mentions a commission to be established under the La Plata River Basin Treaty consisting of representatives from the four states which is to coordinate the compliance with the agreement, concrete steps to establish the commission were lacking (Cassuto and Sampaio 2011; Villar and Ribeiro 2011). Perhaps, the most important obstacle to moving on to more specific and effective regulation and stronger regional institutions is the conflict of interests between different users of the river basin and the lack of consensus. As water is used for many different purposes, this also generates many different interests that have a stake in how regional environmental cooperation takes place. As one

Argentinean researcher noted, water is one of the most important issues for the region, but because of domestic and international conflicts of interests, it is also extremely difficult politically to make binding and specific agreements.[18] For example, the analysis prepared as part of the framework project noted land use changes, in particular the extension of the agricultural frontier due to intensive grain plantations, as an important barrier to the sustainable management of the La Plata basin (CIC 2009, 15–17, 2013) and technical experts have noted that agrochemicals and the pulp industry are important sources of pollution.[19] Civil society organisations working on the Guaraní aquifer have made very similar points as discussed below. Yet, as set out in Chap. 2, powerful domestic and international business interests are clearly in favour of increasing agricultural production. Moreover, governments derive important state income from this sector. This means developing stronger institutions for the environmental protection of the basin's water resources is a difficult and highly political issue which requires not only technical expertise, but also an ability to negotiate between different interests. In this case, the more important conflict of interest is arguably not between different states, but rather between economic and business interests on the one hand and environmental and social concerns on the other.

Furthermore, environmental cooperation in the La Plata basin is very dependent on external funding. While technical experts have frequently mentioned that the lack of funding is a main concern for cooperation in the La Plata basin in general,[20] this is particularly evident in relation to environmental concerns. The three regional development banks in South America, that is the IDB, the Andean Investment Corporation and FONPLATA, are important sources of funding for infrastructure developments in the framework of IIRSA, yet environmental cooperation relies heavily on funding from outside the region, and in particular the GEF. As a consequence, regional environmental cooperation is very much project-based which also makes it vulnerable and fragmented. The dependence on external funding makes long-term planning far more difficult, and there is always the risk that cooperation stops or is delayed when projects finish or when there are disagreements between donors and recipients. Dependence on external funding has also meant that donors have had a significant influence in terms of determining where cooperation takes place and on which topics. Although other agencies have also supported individual initiatives and of course the La Plata basin countries themselves also contribute their share of co-funding, the GEF emerged as the most important funder. From the start, the question of what kinds of environmental problems should be addressed under the GEF has been subject to many debates and marked by serious North South disagreements (Fairman 1996; Gupta 1995; Streck 2001). Controversially, one of the main objectives of the GEF is to support projects of global rather than local importance. As the description of the Guaraní aquifer project shows, this meant that the GEF turned down project proposals for parts of the La Plata basin that it did not consider important enough. The same project also shows the GEF's impact on the contents of projects as it did not accept a project that was based only on research. As the single most important funder, the GEF has thus had a significant impact on regional environmental cooperation, determining what kind of projects and which parts of the La Plata basin receive funding to the detriment of more remote areas.

At the same time, the way the projects have been implemented has not been uncontested. Specifically, in relation to the Guaraní aquifer project, some researchers from public universities criticised the decision to employ private consultancies instead of public universities to carry out the project (Guterres 2009, 38; Novoa 2009, 43). Although the project included a fund for public universities, researchers have pointed out that this only made up a very small proportion of the total budget (Augé 2009, 18). However, as one interviewee also pointed out, individual researchers nevertheless continued working for the project as they were contracted by the consultancies.[21] For several of the projects on rivers, the same happened and researchers or people who had previously worked at a university became involved in the projects.[22] Researchers from the region have therefore been involved in most projects, and this is also crucial in terms of providing continuity, but there are disagreements as to whether projects like these should be carried out by public or private actors. Cooperation between endogenous and exogenous drivers is an important element in this case study, but the role of the different drivers has been scrutinised more and contested more compared to the case study on migratory species examined in the following chapter.

The project on the Pilcomayo River which was the only one that received support from another funder, the EU, makes for an interesting comparison with the GEF. With the EU's priorities being rather different from those of the GEF, the project developed very differently. This only underlines the influence of external funders on regional environmental cooperation in the La Plata basin. In the 1990s, the Trinational Commission for the Development of the Pilcomayo River Basin consisting of all three countries sharing the Pilcomayo River was created to promote the integrated development of the basin (del Castillo Laborde 1999, 192–193, 2008, 285–286). The creation of this commission reflects a rapprochement between the three countries and the shared desire to improve the quality of life of the people living in the basin while at the same time conserving the environment. From the start, this process was supported by the EU in an effort to promote regional integration, and in 2000, the Trinational Commission asked for technical and financial support from the EU resulting in a project which ran from 2002 to 2008. The EU first emphasised the importance of participation of local communities, but now the technical staff working on the issue also believe that this is an important element of river management.[23] Consequently, participation mechanisms have become more institutionalised in the Pilcomayo case than elsewhere in the basin.

However, the experience of the Pilcomayo River has not had a large impact on the La Plata basin regime as a whole. On the one hand, there is relatively little communication, and exchange of information between the experiences in different parts of the basin and the Pilcomayo River is relatively far from the political centres in the La Plata basin. Moreover, the Trinational Commission of the Pilcomayo River still faces considerable challenges. Most importantly, it does not have its own financing mechanism. Instead, the budget varies according to the social and economic reality of the countries, and funding is not assured beyond the very short-term future. In addition, the participatory approach, although valuable, also brings challenges. More and different interests need to be taken into account, and fair representation has to be ensured. Moreover, governments are not familiar with community participation.[24] Taken together, these different elements clearly demonstrate that although environmental cooperation in the La Plata basin has increased significantly since the early 1990s and made some important achievements, it is still a marginal phenomenon which is overshadowed by other objectives and highly dependent on external funding.

This has been exacerbated by the fact that although regional civil society networks with an interest in the socio-environmental concerns of the basin exist and these could constitute an important potential endogenous driver for environmental cooperation, they have been sidelined in the governance of the basin. Although most projects include some components for civil society engagement, this has not translated into notable influence in decision-making. Regional civil society networks became particularly active in relation to the Guaraní aquifer and built links with members of the

Mercosur Parliament, but access for civil society was made more difficult when governments decided that the governance of the aquifer should be addressed by the more technical forums of the La Plata basin regime rather than Mercosur which at least initially had the potential to be more accessible.

The recognition that the four countries share an important aquifer led to important debates and workshops among civil society organisations and some Mercosur institutions. Social movements and networks defending the right to water and denouncing privatisation of water have used the debate on the Guaraní aquifer as a context to bring forward different visions of development and regional integration and linking the debate on the Guaraní aquifer to questions of environmental and social justice. Many of these ideas were reflected in joint declarations resulting from civil society meetings. Organisations have also drawn attention to land use changes that have already caused significant deterioration of surface waters and ecosystems and are likely to present a threat to the aquifer in the future. These include intensive agriculture, the expanding soybean monocultures, pine and eucalyptus and sugar cane plantations. In relation to these, civil society organisations have discussed not only the environmental impact, but also social problems such as forced displacement of people and effects on human health for example from the unregulated use of agrochemicals (Celiberti and Taks 2009; Iglesias and Taks 2009). It is notable that most of these debates have taken place in the same time period, but outside the GEF project with few possibilities of interaction with the project. This has led to criticism that the Guaraní aquifer project did not take civil society into account enough (Guterres 2009, 38; Segovia 2009, 90), as well as demands for a more accessible information system on the aquifer, expressed for example in the conclusions of an international conference on the management of the Guaraní aquifer (CIGSAG 2011).

Civil society organisations also established links with the Mercosur Parliament, also called the Parlasur. With regard to the Guaraní aquifer, the Parliament was one of the first to suggest a joint agreement. Moreover, it proposed creating a commission to examine and compare the water resource legislation of all four countries in detail in order to recommend changes to the national governments to be able to protect the aquifer. It also suggested the creation of a regional research institute, but neither of these were implemented (Villar 2010, 2–3; Villar and Ribeiro 2011, 651). The debate on the Guaraní aquifer in the Mercosur Parliament was led in particular by Dr. Florisvaldo Rosinha, a Brazilian member of the Parliament who was one of the first presidents of the Mercosur Parliament and an enthusiastic supporter of the Mercosur Parliament. Rosinha already had links to researchers and environmental groups working on the Guaraní aquifer in his home town, and he used the space that was created with the establishment of the Mercosur Parliament in order to take the topic forward to a bigger audience at the regional level.[25] Even though the Mercosur Parliament is relatively weak (Malamud and Dri 2013), this provided greater visibility of civil society debates on the aquifer in the region.

However, overlapping mandates and institutional rivalries between Mercosur and the La Plata basin institutions were eventually resolved by governments in favour of the La Plata basin regime. This also had consequences for civil society participation in relation to the aquifer. The creation of the regional organisation Mercosur in 1991 threatened to make the La Plata basin institutions redundant. However, in 2001 the foreign affairs ministers decided to reform and strengthen the CIC in addition to Mercosur rather than abolishing it (Kempkey et al. 2009, 263; Pochat 2011, 502). Nevertheless, since then tasks and responsibilities between the two organisations have not always been very clearly defined, and this has led to confusion and institutional rivalry. This was most evident in relation to the governance of the Guaraní aquifer. As the countries sharing the aquifer corresponded exactly to the Mercosur member countries, several Mercosur institutions expressed interest in the Guaraní aquifer. Yet, Mercosur's environmental institutions, the Mercosur working subgroup on environment SGT6 as well as the Mercosur Meeting of Environment Ministers were never involved in the work on the Guaraní aquifer or the GEF project despite the fact that they had asked to be informed (Hochstetler 2011, 137; Mercosur 2008, 18; Moreno 2011, 73; Villar and Ribeiro 2011, 651). In the end, the GEF project was implemented largely separate from Mercosur structures, and Mercosur was sidelined in the agreement that was signed in 2010 (Villar and Ribeiro 2011, 656). This means Mercosur does not address some of the region's most important transboundary environmental issues, which clearly weakens the regional organisation as a framework for environmental cooperation as discussed in the previous chapter. In addition, the decision of the Southern Cone governments to deal with the governance of the aquifer in the framework of the La Plata basin regime and the CIC rather than Mercosur also limited the possibilities for citizen involvement. As one interviewee explained, the

Mercosur Parliament is more accessible to civil society than the technical forums of the La Plata basin regime:

> And the advantage of Parlasur, which the NGOs appreciated, was that it was a political space and not just a technical one, like for example the projects of the maritime front and the project of the La Plata basin. So for the NGOs, it is easier to try to get involved in Parlasur than in a project of a scientific-technical character.[26]

Similarly, another interviewee noted that the CIC meetings are closed so that civil society activity mostly concentrates on lobbying the national representatives in the CIC.[27] This contrasts with Mercosur where at least formally some meetings are open to registered civil society participants even though in practice there are also many barriers as noted in the previous chapter. Finally, although the GEF projects generally include a component for education, increasing public awareness and participation of stakeholders, this rarely translates into any real influence over decisions that are taken. An interviewee who had worked for an NGO that had carried out some activities as part of the GEF project on the Guaraní aquifer as well as the FREPLATA project, for example, explained that for both projects there was some funding for civil society seminars and workshops, but there was very little civil society participation in the decision-making.[28]

Finally, the GEF's significant involvement in the basin also sparked public criticism and suspicion due to its links to the World Bank which had become extremely unpopular in the region following its promotion of structural adjustment packages during the 1980s and 1990s. Distrust and criticism of the World Bank of course existed already prior to the GEF projects in the La Plata basin, but the lack of openness towards civil society presumably reinforced this. A growing awareness of water scarcity at the global level together with earlier pressure on the Southern Cone states by the World Bank and the IMF for an economic transformation, including privatisation of water management, led to suspicions of plans for foreign dominance over the aquifer (Villar 2007, 69). In this context, several rumours started to circulate according to which foreign companies buy land above the aquifer in order to exploit the water if water becomes scarce globally (Sindico 2011, 262). This distrust of the main funder also made cooperation more difficult. An interviewee who had worked for an Argentinean NGO which received a small part of the project funding in order to carry out activities involving civil society, for example, explained that initially it was very difficult to convince people to come along to events organised by the NGO regarding the Guaraní aquifer. Although the NGO was Argentinean with local staff, because the overall project on the aquifer was funded by the GEF, which is associated with the World Bank, many people were sceptical. The NGO therefore first had to convince participants that the discussion would be open and that there was no imposition of the World Bank's agenda.[29] Overall, it is evident that there is an interest on the part of civil society in relation to socio-environmental concerns in the La Plata basin. However, regional civil society initiatives have struggled to become a key driver for regional environmental cooperation as governments have given preference to regional institutions focussing on technical questions rather than the wider social and political concerns expressed by civil society.

Overall, robust forms of regional environmental cooperation linking formal agreements and commitments on the part of governments with regular joint activities in practice have therefore developed in the framework of the La Plata basin regime since the 1990s. The crucial factors in this process were more favourable domestic and international context conditions as well as different endogenous drivers, notably networks of researchers and technical staff working together with external funders and in particular the GEF towards specific initiatives of environmental cooperation. Yet, other important potential endogenous drivers from civil society have largely been sidelined in the process as governments have given preference to the La Plata basin regime over Mercosur and opportunities for civil society participation in cooperation projects have been relatively limited. This could be read as an attempt at depoliticising the governance of water which has been a highly contentious issue in the region, as demonstrated on several occasions, including the pulp mill conflict between Argentina and Uruguay, the various hydropower projects in the region and the Cochabamba water wars in Bolivia noted in Chap. 2. As a result, civil society initiatives have been more influential in stopping or downscaling specific infrastructure projects in the basin than in shaping regional environmental cooperation. Environmental cooperation in the framework of the La Plata basin regime is thus an example of cooperation on an issue which has received a high level of public attention and which is a politically sensitive issue. This is quite different from the following case study on the protection of migratory species which has had much less visibility and public attention.

NOTES

1. The actual electricity production of a hydropower station depends on water availability and rainfall and is therefore susceptible to climate change. In 2014 when Brazil faced severe droughts energy production fell and the Three Gorges Dam produced more energy in that year, but in 2015 Itaipú regained its yearly electricity production leading position with 89.2 million MWh (Itaipu Binacional 2016).

2. For examples of rivalries between the basin countries and military operations in relation to the rivers of the La Plata basin see: Elhance (1999), Kempkey et al. (2009), Da Rosa (1983).

3. Interviews, Comisión Mixta Argentino Paraguaya del Rió Paraná (COMIP), Buenos Aires, 2011; Comisión Mixta Paraguayo - Argentino del Rió Paraná (COMIP), Asuncion, 2011.

4. Interview, Comisión Mixta Paraguayo - Argentino del Rió Paraná (COMIP), Asuncion, 2011, author's translation.

5. Interview, Iglesia Evangélica del Río de la Plata, Buenos Aires, 2010, author's translation.

6. Interview, Sobrevivencia/Amigos de la Tierra Paraguay, Asuncion, 2011.

7. Debates regarding hydropower projects are not limited to the La Plata river basin, but take place across Latin America. Well known cases include the protests of indigenous and environmental groups against the Belo Monte dam in the Brazilian Amazon or the HydroAysén project in Chilean Patagonia which was cancelled by the government in 2014 following severe public pressure (Edwards and Roberts 2015, 52–57).

8. For an overview see: del Castillo Laborde (2008, 284–286), Pochat (2011, 505–507).

9. Technical staff explained that information on most topics is publicly available. However, water quality data is a sensitive issue which requires authorisation, but if requested this is usually granted (interviews Comisión Trinacional Para el Desarrollo de la Cuenca del Río Pilcomayo – Dirección Ejecutiva, Formosa, 2011).

10. Interview, Ministerio de Vivienda, Ordenamiento Territorial y Medio Ambiente, Dirección Nacional de Medio Ambiente, Montevideo, 2011; See also: Celiberti and Taks (2009), Iglesias and Taks (2009), Villar (2007, 67–68).

11. Interview, Ministerio de Vivienda, Ordenamiento Territorial y Medio Ambiente, Dirección Nacional de Medio Ambiente, Montevideo, 2011.

12. Interview, Ministerio de Vivienda, Ordenamiento Territorial y Medio Ambiente, Dirección Nacional de Agua, Montevideo, 2011, author's translation.

13. Interview, Ministerio de Vivienda, Ordenamiento Territorial y Medio Ambiente, Dirección Nacional de Agua, Montevideo, 2011.
14. Interview, UNDP, Montevideo, 2010, author's translation.
15. Interviews Comisión Trinacional Para el Desarrollo de la Cuenca del Río Pilcomayo – Dirección Ejecutiva, Formosa, 2011.
16. Interview Ministerio de Vivienda, Ordenamiento Territorial y Medio Ambiente, Dirección Nacional de Medio Ambiente, Montevideo, 2011.
17. Interview, Proyecto FREPLATA II, Montevideo, 2011.
18. Interview, Consejo Nacional del Investigaciones Científicas y Técnicas (CONICET), Rosario, 2011.
19. Interviews, Comisión Mixta Argentino Paraguaya del Rió Paraná (COMIP), Buenos Aires, 2011; Comisión Mixta Paraguayo - Argentino del Rió Paraná (COMIP), Asuncion, 2011.
20. Interviews, Comisión Mixta Paraguayo - Argentino del Rió Paraná (COMIP), Asuncion, 2011; Proyecto FREPLATA II, Montevideo, 2011; Comisión Trinacional Para el Desarrollo de la Cuenca del Río Pilcomayo – Dirección Ejecutiva, Formosa, 2011; see also: del Castillo Laborde (1999, 198), Pochat (1999, 144, 2011, 507).
21. Interview, Universidad de la República, Montevideo, 2011.
22. Interviews, Proyecto FREPLATA II, Montevideo, 2011; UNDP, Montevideo, 2010.
23. Interviews Comisión Trinacional Para el Desarrollo de la Cuenca del Río Pilcomayo – Dirección Ejecutiva, Formosa, 2011.
24. Interviews Comisión Trinacional Para el Desarrollo de la Cuenca del Río Pilcomayo – Dirección Ejecutiva, Formosa, 2011.
25. Interview, Universidad de la República, 2011; see also Rosinha (2009).
26. Interview, Universidad de la República, 2011, author's translation.
27. Interview, Sobrevivencia/Amigos de la Tierra Paraguay, Asuncion, 2011.
28. Interview, former NGO staff member, Centro Tecnológico para la Sustentabilidad, Buenos Aires, 2011.
29. Interview, former NGO staff member, Centro Tecnológico para la Sustentabilidad, Buenos Aires, 2011.

References

Augé, Miguel. 2009. Acuífero Guaraní. Características Hidrogeológicas Y Gestión Para Su Manejo. In *Acuífero Guaraní, Por Una Gestión Participativa. Voces Y Propuestas Desde El Movimiento Del Agua*, ed. Verónica Iglesias and Javier Taks. Montevideo: Casa Bertolt Brecht.

Bartolome, Leopoldo J., and Christine M. Danklmaier. 2012. Hydrodevelopment and Population Displacement in Argentina. In *Impacts of Large Dams: A Global*

Assessment, ed. Cecilia Tortajada, Dogan Altinbilek, and Asit K. Biswas. Berlin: Springer.

Bucher, Enrique H., and Paul C. Huszar. 1995. Critical Environmental Costs of the Paraguay-Paraná Waterway Project in South America. *Ecological Economics* (15): 3–9.

Bueno, María del Pilar. 2010. *De Estocolmo a La Haya – La Desarticulación de Las Políticas Ambientales En La Argentina*. Rosario: UNR Editora – Editorial de la Universidad Nacional de Rosario.

Cassuto, David N., and Romulo S.R. Sampaio. 2011. Keeping It Legal: Transboundary Management Challenges Facing Brazil and the Guarani. *Water International* 36 (5): 661–670.

Celiberti, Lilian, and Javier Taks (eds.). 2009. *El Acuífero Guaraní En Debate*. Montevideo: Cotidiano Mujer.

CIC. 2009. Project Document—Sustainable Management of the Water Resources of the La Plata Basin with Respect to the Effects of Climate Variability and Change. Comité Intergubernamental Coordinador de los Países de la Cuenca del Plata (CIC). http://www.cicplata.org.

CIC. 2013. Programa Marco Para La Gestión Sostenible de Los Recursos Hídricos de La Cuenca Del Plata, En Relación Con Los Efectos de La Variabilidad Y El Cambio Climático. http://www.proyectoscic.org.

CIGSAG. 2011. *Carta de São Paulo*. São Paulo: Conferência Internacional - A Gestão do Sistema Aquífero Guarani (CIGSAG). http://www.isarm.org/publications/402.

Da Rosa, J. Eliseo. 1983. Economics, Politics and Hydroelectric Power: The Parana River Basin. *Latin American Research Review* 18 (3): 77–107.

Del Castillo Laborde, Lilian. 1999. The Plata Basin Institutional Framework. In *Management of Latin American Rivers Basins: Amazon, Plata, and São Francisco*, ed. Asit K. Biswas, Newton V. Cordeiro, Benedito P.F. Braga, and Cecilia Tortajada. Tokyo: United Nations University Press.

Del Castillo Laborde, Lilian. 2008. The Rio de La Plata River Basin: The Path Towards Basin Institutions. In *Management of Transboundary Rivers and Lakes*, ed. Olli Varis, Cecilia Tortajada, and Asit K. Biswas. Berlin: Springer.

Dudek, Carolyn M. 2013. Transmitting Environmentalism?: The Unintended Global Consequences of European Union Environmental Policies. *Global Environmental Politics* 13 (2): 109–127.

Edwards, Guy, and J. Timmons Roberts. 2015. *A Fragmented Continent—Latin America and the Global Politics of Climate Change*. Cambridge, MA: MIT Press.

Elhance, Arun P. 1999. *Hydropolitics in the Third World—Conflict and Cooperation in International River Basins*. Washington, DC: United States Institute of Peace Press.

Fairman, David. 1996. The Global Environment Facility: Haunted by the Shadow of the Future. In *Institutions for Environmental Aid*, ed. Robert O. Keohane and Marc A. Levy. London: MIT Press.

FREPLATA. 2005. *Análisis Diagnostico Transfronterizo Del Rio de La Plata Y Su Frente Marítimo, Documento Técnico*. Montevideo. http://www.freplata.org/documentos/adt/default.asp.

GEF. 2009. Request for CEO Endorsement/Approval—Project Title: Reducing and Preventing Land-Based Pollution in the Rio de La Plata/Maritime Front through Implementation of the FREPLATA Strategic Action Programme. http://www.undp.org.uy/showProgram.asp?tfProgram=185.

Gilman, Patrick, Víctor Pochat, and Ariel Dinar. 2008. Whither La Plata? Assessing the State of Transboundary Water Resource Cooperation in the Basin. *Natural Resources Forum* 32 (August): 203–214.

Giraut, M.A., C. Laboranti, C. Magnani, and L. Borello. 2010. Guarani Aquifer System Project: Strengths and Weaknesses of Its Implementation. In *Transboundary Aquifers: Challenges and New Directions. ISARM2010 International Conference, 6–8 December, UNESCO, Paris. Pre-Proceedings*. Paris: UNESCO. http://www.isarm.org/publications/397.

Gottgens, Johan F., James E. Perry, Ronald H. Fortney, Jill E. Meyer, Michael Benedict, and Brian E. Rood. 2001. The Paraguay-Paraná Hidrovía: Protecting the Pantanal with Lessons from the Past. *BioScience* 51 (4): 301–308.

Gupta, Joyeeta. 1995. The Global Environment Facility in Its North-South Context. *Environmental Politics* 4 (1): 19–43.

Guterres, José Augusto. 2009. ¿Qué Sabemos Del Acuífero Guaraní Y La Gestión de Su Conocimiento? In *Acuífero Guaraní, Por Una Gestión Participativa. Voces Y Propuestas Desde El Movimiento Del Agua*, ed. Verónica Iglesias and Javier Taks. Montevideo: Casa Bertolt Brecht.

Hochstetler, Kathryn. 2002. After the Boomerang: Environmental Movements and Politics in the La Plata River Basin. *Global Environmental Politics* 2 (4): 35–57.

Hochstetler. 2011. Under Construction—Debating the Region in South America. In *Comparative Environmental Regionalism*, ed. Lorraine Elliott and Shaun Breslin. Oxon: Routledge.

Iglesias, Verónica, and Javier Taks (eds.). 2009. *Acuífero Guaraní, Por Una Gestión Participativa. Voces Y Propuestas Desde El Movimiento Del Agua*. Montevideo: Casa Bertolt Brecht.

Itaipu Binacional. 2016. Itaipu Binacional. https://www.itaipu.gov.br.

Kempkey, Natalie, Margaret Pinard, Víctor Pochat, and Ariel Dinar. 2009. Negotiations Over Water and Other Natural Resources in the La Plata River Basin: A Model for Other Transboundary Basins? *International Negotiation* 14 (May): 253–279.

Malamud, Andrés, and Clarissa Dri. 2013. Spillover Effects and Supranational Parliaments: The Case of Mercosur. *Journal of Iberian and Latin American Research* 19 (2): 224–238.

Mercosur. 2008. *La Temática Ambiental En El Mercosur: Evolución Y Perspectivas.* Rio de Janeiro.

Molle, François. 2009. River-Basin Planning and Management: The Social Life of a Concept. *Geoforum* (40) (May): 484–494.

Moreno, Alicia. 2011. La Necesidad de Una Estrategia Ambiental En El MERCOSUR. *Densidades* (6): 63–77.

Novoa, Luis Fernando. 2009. Agua Para El Molino Del Capital: Infraestructura Hídrica Y Acuífero Guaraní. In *Acuífero Guaraní, Por Una Gestión Participativa. Voces Y Propuestas Desde El Movimiento Del Agua*, ed. Verónica Iglesias and Javier Taks. Montevideo: Casa Bertolt Brecht.

O'Shaughnessy, Hugh, and Edgar Venerando Ruiz Díaz. 2009. *The Priest of Paraguay—Fernando Lugo and the Making of a Nation.* London: Zed Books.

OAS. 2005a. Pantanal and the Upper Paraguay River Basin—Implementation of Integrated Watershed Management Practices for the Pantanal and the Upper Paraguay River Basin. OAS. http://www.oas.org/en/sedi/dsd/IWRM/Past_Projects/Pantanal/project_default.asp.

OAS. 2005b. Guarani Aquifer System—Environmental Protection and Sustainable Development of the Guarani Aquifer System. OAS. http://www.oas.org/DSD/Events/english/Documents/OSDE_7Guarani.pdf.

OAS. 2005c. La Plata River Basin—A Framework for the Sustainable Management of Its Water Resources with Respect to the Hydrological Effects of Climatic Variability and Change. OAS. http://www.oas.org/en/sedi/dsd/IWRM/Past_Projects/La_Plata/project_default.asp.

Pochat, Víctor. 1999. Water-Resources Management of the Plata Basin. In *Management of Latin American Rivers Basins: Amazon, Plata, and São Francisco*, ed. Asit K. Biswas, Newton V. Cordeiro, Benedito P.F. Braga, and Cecilia Tortajada. Tokyo: United Nations University Press.

Pochat, Víctor. 2011. International Agreements, Institutions and Projects in La Plata River Basin. *International Journal of Water Resources Development* 27 (3): 497–510.

Proyecto de Gestión Integrada y Plan Maestro de la Cuenca del Río Pilcomayo. 2008. *Proyecto de Gestión Integrada Y Plan Maestro de La Cuenca Del Río Pilcomayo.* Tarija.

Rosinha, Florisvaldo. 2009. Acüífero Guaraní: Inércia E Destruição. In *El Acuífero Guaraní En Debate*, ed. Lilian Celiberti, and Javier Taks. Montevideo: Cotidiano Mujer.

Segovia, Diego. 2009. Un Acuífero Guaraní Sin Guaraníes. In *El Acuífero Guaraní En Debate*, ed. Lilian Celiberti, and Javier Taks, 83–91. Montevideo: Cotidiano Mujer.

Sindico, Francesco. 2011. The Guarani Aquifer System and the International Law of Transboundary Aquifers. *International Community Law Review* (13): 255–272.

Sindico, Francesco. 2016 Past, Present and Future of the Salto/Concordia Guarani Aquifer Binational Commission. In *Working Paper No.4, 2016 Groundwater Governance: Drawing Connections between Science, Knowledge and Policy-Making*, ed. Francesco Sindico and Alberto Manganelli. Strathclyde Centre for Environmental Law and Governance.

Streck, Charlotte. 2001. The Global Environment Facility—A Role Model for International Governance? *Global Environmental Politics* 1 (2): 71–94.

Tratado de La Cuenca Del Plata. 1969. http://www.cicplata.org/?id=tratado.

Tucci, Carlos E.M., and Robin T. Clarke. 1998. Environmental Issues in the La Plata Basin. *Water Resources Development* 14 (2): 157–173.

Tussie, Diana, and Patricia Vásquez. 2000. Regional Integration and Building Blocks: The Case of Mercosur. In *The Environment and International Trade Negotiations—Developing Country Stakes*, ed. Diana Tussie. Houndmills: Macmillan Press Ltd.

Uhlin, Anders. 2011. National Democratization Theory and Global Governance: Civil Society and the Liberalization of the Asian Development Bank. *Democratization* 18 (3): 847–871.

Villar, Pilar Carolina. 2007. A Gestão Internacional Dos Recursos Hídricos Subterrâneos Transfronteiriços E O Aqüífero Guarani. *REGA - Revista de Gestão de Água Da América Latina* 4 (1): 63–74.

Villar, Pilar Carolina. 2010. Moving Toward Managing the Guarani Aquifer: The Brazilian Case. In *Transboundary Aquifers: Challenges and New Directions. ISARM2010 International Conference, 6–8 December, UNESCO, Paris. Pre-Proceedings*. Paris: UNESCO. http://www.isarm.org/publications/397.

Villar, Pilar Carolina. 2016. International cooperation on transboundary aquifers in South America and the Guarani Aquifer case. *Revista Brasileira de Política Internacional* 59 (1): e007.

Villar, Pilar Carolina, and Wagner Costa Ribeiro. 2011. The Agreement on the Guarani Aquifer: A New Paradigm for Transboundary Groundwater Management? *Water International* 36 (5): 646–660.

Vogler, John. 2007. The International Politics of Sustainable Development. In *Handbook of Sustainable Development*, ed. Giles Atkinson, Simon Dietz, and Eric Neumayer. Cheltenham: Edward Elgar.

Waisbord, Silvio, and Enrique Peruzzotti. 2009. The Environmental Story That Wasn't: Advocacy, Journalism and the Asambleismo Movement in Argentina. *Media, Culture and Society* 31 (5): 691–709.

WWF. 2007. *World's Top 10 Rivers at Risk*. http://wwf.panda.org/about_our_earth/about_freshwater/rivers/.

CHAPTER 5

Species Protection at the Regional Level: The Convention on Migratory Species in the Southern Cone

Regional environmental cooperation on endangered species developed very differently from both, the previous case study and the initiatives of regional environmental cooperation in Mercosur discussed in Chap. 3. Most importantly, it evolved separately from other economic and political regional integration processes and regional institutions. Whereas in the other case study the institutional framework is provided by a regional resource regime which was created for political and economic reasons and where environmental concerns are only one among several objectives, in this case study a global environmental regime targeting a specific environmental concern serves as a framework. This also means that the scope of issues addressed is narrower and focuses specifically on the protection of endangered migratory species. This chapter first introduces the Convention on Migratory Species and then examines the first steps towards regional environmental cooperation which were initiated by endogenous drivers, notably networks of researchers and conservation NGOs. Initially, research was important because it led to the realisation that certain groups of species are endangered and that they regularly migrate across national boundaries. This established the scope of the issue and defined the boundaries for cooperation in ecological terms, that is the countries sharing the habitat for particular species. Regional environmental cooperation on migratory species therefore clearly started from cooperation in practice in the form of regular meetings and exchanges of information, as well as joint research and

© The Author(s) 2017
K.M. Siegel, *Regional Environmental Cooperation in South America*,
International Political Economy Series, DOI 10.1057/978-1-137-55874-9_5

conservation activities between researchers, conservation NGOs and, in some cases, national park administrations in neighbouring countries. During the 2000s the CMS started to turn into the main institutional framework for the protection of migratory species in the region. This was actively promoted by the regional conservation networks that had formed earlier, primarily because the CMS offered a way of strengthening formal cooperation between governments which had been only weakly developed until then. Moreover, regional conservation networks benefitted from the relatively open institutional framework of the CMS which provides important possibilities for the involvement of professional conservation NGOs and also opens up ways to access national governments. Simultaneously, the CMS Secretariat itself turned its attention towards the Southern Cone region and worked together with non-state actors in the region in order to promote the convention with national governments. As a result, a global environmental regime that had already existed for some time became a framework for regional environmental cooperation in South America.

Overall, the links between the endogenous and exogenous drivers promoting regional environmental cooperation are relatively strong in this case study, making this a comparatively robust example of cooperation. Moreover, although the activities that make up cooperation in practice are also heavily dependent on external funding, this comes from a variety of different sources. Consequently, cooperation is less dependent on one single funder and its characteristics are less shaped by any particular donor. At the same time, it is important to note that this comparatively higher level of robustness is linked to very specific conditions. Most importantly, as examined in the last section of the chapter, the endogenous drivers involved in this case study are professional conservation NGOs which have significant resources to offer to governments and have approached the environmental issue at stake mostly from a technical perspective that does not raise more contested questions of environmental justice or rights. Moreover, despite these achievements, regional cooperation on the protection of migratory species is also marked by marginality. This is evident in the non-binding nature of the agreements that have been signed and the high dependence of governments on professional NGOs with autonomous sources of funding for their implementation.

The Convention on Migratory Species

The Convention on Migratory Species, sometimes also called the Bonn Convention, is part of the broader group of global environmental treaties. These address environmental concerns which are seen as globally important and they are open to all countries. The concern for migratory species was taken up at the global level at the Stockholm Conference on the Human Environment in 1972 due to serious concerns regarding the significant loss of some migratory species because of excessive hunting, destruction of habitat and contamination of feeding grounds. Consequently, the Federal Republic of Germany started to draft a convention which was concluded in 1979 and entered into force in 1983 (Caddell 2005, 114). Migratory species is an issue that per se requires transboundary cooperation, at least according to its definition by the CMS. The CMS thus states that "'Migratory species' means the entire population or any geographically separate part of the population of any species (…), a significant proportion of whose members cyclically and predictably cross one or more national jurisdictional boundaries" (CMS 1979: Article I). With this definition, the CMS excludes species whose migratory range falls within one country only (de Klemm 1994, 70). Migratory species are particularly vulnerable because they often travel huge distances and cross one or more political boundaries. As a consequence, it is not sufficient to protect them and conserve their habitat in one country, but a minimum of conservation and protection along their whole migratory route is necessary in order to conserve the species (Caddell 2005, 113–114). Similar to other global environmental conventions, the most important decision-making forum is the so-called Conference of the Parties which brings together representatives of all the member states and takes place every two to three years. In addition, the CMS includes a secretariat to provide administrative support on a daily basis. The secretariat has no formal decision-making powers, but can make recommendations or shape implementation through its daily work.

Whereas in the previous case study an existing regional regime serves as the institutional framework for environmental cooperation in the Southern Cone, the link to the regional level is less obvious in the case of a global environmental convention. Nevertheless, global environmental conventions can also play a role in regional environmental cooperation. Governments may thus choose to reinforce the implementation of a global environmental convention at the regional level because this promises to be more practical or effective. Regional arrangements under the umbrella of a global environmental convention can also help to address regional differences and deal with specific issues of particular regions more effectively (Bauer 2009; Najam 2004; Selin 2012). In the Southern Cone, regional environmental cooperation also takes place in the framework of other global conventions, such as the UNCCD, but this is less developed than in the case of the CMS. The environmental issues addressed by the CMS, in fact, lend themselves more to being addressed at the regional level even though they may be a concern for countries all over the globe. While the CMS provides a global framework for the protection of migratory species in general, the migratory route of a specific species determines which countries are required to cooperate in order to conserve that specific species. As neighbouring countries have to work together the protection of specific species is therefore much more a regional than a global concern. In fact, the protection of migratory birds was one of the first topics for environmental cooperation between states, and initially, it took place at the regional rather than the global level (Balsiger et al. 2012, 12–13).

The CMS promotes and facilitates agreements for specific species or groups of species through additional agreements under the umbrella of the convention. The most developed of these agreements are legally binding and have their own institutional structure, including a secretariat and regular meetings. This is the case for example for two regional agreements relating to cetaceans or the African-Eurasian Waterbird Agreement (Caddell 2005, 126–134). Other agreements are non-binding, so-called Memoranda of Understanding whose aim is to attain immediate conservation objectives and coordinate measures in relation to administration and scientific research, often in cooperation with relevant NGOs (Caddell 2005; CMS 2006, 3). Agreements should cover the whole range of the species concerned and are therefore open to all range states, including those countries that are not parties to the CMS (CMS 1979, Article V). In addition to these species-specific activities, the CMS also promotes cooperation at the regional level more generally and has, for example, supported several meetings for countries in Latin America and the Caribbean. These have served to provide information about the CMS and exchange experiences on topics that are particularly relevant to this region.

The CMS now has 124 parties and most South American countries are members, with the exception of Colombia, Guyana, Suriname and Venezuela. Chile joined as the first South American country in 1983, the other South American countries joined mostly throughout the 1990s with the exception of Bolivia and Ecuador which joined in 2003 and 2004 respectively and Brazil which joined in 2015 (CMS 2016). Active participation of Southern Cone and some neighbouring countries has increased significantly since 2006 and since then the convention has developed into a framework for cooperation in the region. In the space of only four years (2006–2010) four memoranda of understanding were signed between South American countries. These are the Memorandum of Understanding concerning Conservation Measures for the Ruddy-headed Goose signed between Argentina and Chile in November 2006; the Memorandum of Understanding on the Conservation of Southern South American Migratory Grassland Bird Species and their Habitats between Argentina, Bolivia, Brazil, Paraguay and Uruguay signed in August 2007; the Memorandum of Understanding on the Conservation of High Andean Flamingos and their Habitats signed by Bolivia, Chile and Peru in December 2008; and the Memorandum of Understanding between the Argentine Republic and the Republic of Chile on the Conservation of the Southern Huemul signed in December 2010. This development is remarkable in two respects. First, it represents a significant and unprecedented increase in activity on the part of South American governments in relation to the CMS. Second, the memoranda are quite concentrated in geographic terms and mostly cover migratory species and their habitats in the Southern Cone and some neighbouring countries.

Drivers for Regional Cooperation in the Framework of the CMS

The initial steps towards regional environmental cooperation in the framework of the CMS were largely taken by endogenous drivers, notably networks of researchers and conservation NGOs together with the staff of national parks in some cases. Initially, research was important because it led to the realisation that certain groups of species are endangered and that they regularly migrate across national boundaries. This established the scope of the issue and defined the boundaries for cooperation in ecological terms, that is the countries sharing the habitat for particular species. Regional environmental cooperation on migratory species therefore clearly started from cooperation in practice in the form of regular meetings, exchanges of information and joint research and conservation activities between researchers, conservation NGOs and in some cases also national park administrations in neighbouring countries. After all the Southern

Cone countries with the exception of Brazil had joined the CMS in the 1990s, the global environmental convention gradually turned into the institutional framework for the protection of migratory species during the 2000s. This was actively promoted by the regional conservation networks that had formed earlier, primarily because the CMS offered a way of strengthening formal cooperation between governments which had been only weakly developed until then, and thus achieve more robust cooperation. Regional conservation networks benefitted from the relatively open institutional framework of the CMS which provides possibilities for the involvement of professional NGOs and also opens up ways to access national governments. Simultaneously, the CMS Secretariat itself took the initiative to promote the convention outside Europe and Africa where most of its activities had taken place initially. As part of this, the secretariat actively sought links to Southern Cone governments and worked together with conservation networks in the region to convince governments of the value of the convention. The development of robust forms of regional environmental cooperation on migratory species and in the framework of the CMS was therefore the result of the confluence of two factors; the activities of regional conservation networks on the one hand; and policy changes as well as the initiatives of particular members of staff within the CMS Secretariat which brought the focus to the Southern Cone on the other.

Overall, this case study is characterised by relatively dense links between the different endogenous and exogenous drivers promoting regional environmental cooperation. To a large extent, this is due to the existence of proactive regional conservation networks committed to species protection whose members work for NGOs, universities, national park administrations and environmental agencies of national governments as well as the CMS Secretariat. Moreover, several members of these networks also fit the description of "bilateral activists" (Steinberg 2001, 3–26, 2003) who are well-connected internationally and able to access international sources of funding and expertise, but also know the domestic policy context very well to promote species protection with national governments and other domestic actors. Moreover, although the activities that make up cooperation in practice are also heavily dependent on external funding, this comes from a variety of different sources. Consequently, cooperation is less dependent on one single funder and its characteristics are less shaped by any particular donor. In addition, regional conservation networks have also used a variety of tools and strategies and addressed the issue from different angles. All of these factors mean that regional environmental cooperation is comparatively more robust in this case study. Nevertheless, this comparatively higher level of robustness is only possible under very specific conditions. First, the CMS provides a relatively open institutional framework only to conservation NGOs which are deemed qualified in the protection of migratory species and which have been approved by governments. As a consequence professional NGOs which are willing to work within the structures of the CMS together with governments and which can offer significant expertise and resources to governments are much more likely to be able to link cooperation in practice to formal cooperation than grassroots movements with more limited resources. Second, the protection of migratory species is not a particularly salient or politically sensitive topic.

Although the four memoranda were mostly developed in isolation from each other with little interaction between the different groups working on each agreement, there are significant commonalities in terms of how the memoranda were developed. These include the types of actors involved and their objectives as well as their motivations for working in the framework of the CMS. Moreover, the time frame is very similar. In all four cases cooperation between regional networks of researchers and NGOs as well as some state agencies, notably national park administrations, started to increase significantly from the 1990s onwards. As in the other two case studies, it is thus likely that regional environmental cooperation in the framework of the CMS also benefitted from the more favourable domestic and international context. In the following section I will first outline the origins for each memorandum and then examine the process as a whole in order to explain why regional environmental cooperation developed in this particular framework and in this geographical area.

In the case of the ruddy-headed goose from the late 1990s onwards the NGO *Wetlands International* through its office in Buenos Aires, carried out several projects in order to get more precise information regarding the species, to raise awareness, and carry out concrete protection measures. The different project reports show a clear evolution over the years. While at the beginning the main objectives were to establish where the main breeding, nesting and wintering sites were and how many birds there were, later research focussed on more specific details such as habitat use and the migratory route as well as comparing new figures of abundance with old ones. Throughout the different projects the NGO and their partners also included more people in the awareness raising and education campaigns, from establishing first contacts with authorities and hunting associations to education campaigns, developing and distributing brochures as well as articles in scientific journals (Wetlands International 1998, 2000, 2004, 2009). At the later stages, the NGO worked with authorities in order to update geese hunting regulations in the province of Buenos Aires where the geese spend the winter, elaborating a national conservation plan and carrying out workshops with the National Fauna Direction (Wetlands International 2009, 3).

In the case of the grassland birds memorandum the local branches of an international NGO equally played a crucial role. As set out in Chap. 2, large parts of the Southern Cone have since the 1990s increasingly been turned to intensive agricultural production, and, in particular, soybean. This has radically changed the natural grasslands that characterised this region previously. Many of the species inhabiting these grasslands are therefore facing the loss or degradation of their habitat and are also starting to disappear. This includes several species of migratory grassland birds which live in the Southern Cone part of the year on their migratory routes. The international NGO *BirdLife International* and its local partner organisations became aware of the issue and started to be concerned about the conservation of grassland birds. In the early 2000s, the local partner organisations of *BirdLife International* in the Southern Cone started some activities regarding monitoring and research of grassland birds as well as developing more sustainable cattle ranching models to preserve the grasslands. These activities were led from the *Aves Argentinas* office, the Argentinean partner of *BirdLife International* and initially *Aves Argentinas* also received CMS contributions for this (CMS 2002, 15, 2005a, 15–18). To strengthen this further *BirdLife International* developed the idea of a memorandum of understanding under the umbrella of the CMS and started looking for ways to achieve this. As examined in more detail below, the relationship between their local partner organisation and the CMS was particularly strong in Paraguay because a member of their partner organisation, *Guyra Paraguay*, was nominated as the Scientific Councillor for the CMS by the government. *BirdLife International* therefore chose Paraguay as the entry point to introduce the idea and work towards the development of a memorandum of understanding. The NGO worked closely with the Paraguayan government in order to promote the memorandum.[1]

In the case of the memorandum on Andean flamingos, initial research also played a crucial role in exposing the need for further protection. Until the late 1990s, very little was known about all three species of flamingos living in the high Andes. While there was incomplete information regarding the summer distributions, the movement of the flamingos during the winter was largely hypothetical (Caziani et al. 2007, 277). This only started to change with a series of simultaneous surveys carried out in Argentina, Bolivia, Chile and Peru. The first census took place in 1997 and was then repeated every year until 2000 and then continued at 5 year intervals. The first census was initiated by the Peruvian NGO *Perú Verde* and one of their researchers who got into contact with colleagues in the other countries and organised a first meeting to exchange information. Two months later the first simultaneous census was held. Both the census and the preparatory meeting were supported by the Wildlife Conservation Society.[2]

The first exchanges also resulted in the creation of the *Grupo de Conservación Flamencos Altoandinos* (High Andes Flamingo Conservation Group or GCFA). The GCFA is an international working group consisting of scientists and specialists of conservation and protected areas of Argentina, Bolivia, Chile and Peru coming from the public sector, civil society and the private sector. It consists of a permanent council and secretariat which changes every two years (Marconi 2010, 3, 37). While the main objectives of the first censuses were to gather basic data on abundance and distribution, the researchers carrying out the work soon also became aware of problems regarding the conservation of these species. A researcher described the evolution of the work like this:

> It was a group of researchers and we worked on the same topic and we had meetings to see what we're working on and to have a regional view of the situation of these species. And the first gap that we found was that we didn't know how many there were and where they were. That was the first big question. Once we knew how many there were and where they were, we thought about which strategy to use to get a strategy for conservation, but always at the regional level. Because with these species it doesn't make sense to work at the level of the countries because they don't work according to political borders. Because, even on a daily basis, groups that are nesting in Bolivia get food in Argentina, they come and go in one day. So international participation is necessary.[3]

As a result, the objective of the GCFA is not only to improve knowledge and scientific research on the flamingos of the high Andes, but also to promote the active participation of local communities, the development and implemention of management plans, awareness raising and the creation of protected areas (GCFA 2011). The group thus has two areas of work, research on the one hand and administration and political issues on the other.[4]

The last memorandum that has been signed so far, in 2010, addresses the protection of the huemul, a species of deer living in the Southern Andes. The two range states, Argentina and Chile, have been cooperating on the issue already for two decades. Since 1992 bilateral technical meetings have been held on a regular basis. These meetings were attended by staff from official institutions and NGO representatives from both countries and they resulted in recommendations on various topics, including legislation, research, management, education and conservation (Serret 2001, 104). This has also resulted in cooperation on the ground, for example with park rangers participating in activities in national parks of the neighbouring country (Corporación Nacional Forestal 2010; Secretaría de Ambiente y Desarrollo Sustentable de la Nación—Argentina 2002, 14; Serret 2001, 110). Overall, the national park administrations have therefore been one of the main driving forces and they play a crucial role in hosting meetings, organising education and awareness-raising campaigns as well as collaborating with research (Corporación Nacional Forestal 2010, 2011a, b; Secretaría de Ambiente y Desarrollo Sustentable de la Nación—Argentina 2008, 2009; Serret 2001, 103, 109–111; Vila et al. 2006). In addition to the national park administrations, NGOs and private foundations have also been involved in research and conservation activities and campaigns (Comité Nacional Pro Defensa de la Flora y Fauna Chile 2013; Huilo Huilo Foundation2013).

The trajectories of the four memoranda clearly show several commonalities although there has been no notable interaction between actors working on the different agreements. First, in all memoranda except for the last one on the huemul, the extent of the threat to the different species mostly became evident in the 1990s. In these cases, researchers played a fundamental role in two respects. Research was the basis for realising that the species were endangered and it also helped understand the distribution and behaviour of the different animals better. The latter was crucial for realising that the animals in question regularly cross national borders which in turn established them as species that fall under the remit of the CMS. The case of the huemul is different, perhaps because both the threat to the survival of the South Andean deer and awareness of the need for protection, go back longer. The abundance and distribution of the huemul has been declining since the arrival of the European colonisers, but some form of legal protection of the huemul also dates back as far as the 1930s (Vila et al. 2006, 263). However, in the other cases research on the extent of the problem and its transboundary dimensions was crucial as a first step. Following on from this, researchers have then also taken the initiative to promote the protection of the different species. Regional environmental cooperation on the protection of migratory species therefore clearly started from cooperation in practice and joint activities of networks including researchers, NGOs and staff working for national parks were well developed before the Southern Cone governments signed the different memoranda of understanding. The next section outlines the processes that led to the establishment of formal cooperation in the form of agreements between governments and examines why it was the CMS that became the institutional framework for regional environmental cooperation.

The CMS as a framework for regional environmental cooperation on migratory species in the Southern Cone has been promoted from two angles. First, regional conservation networks have been concerned with strengthening cooperation and making it more robust. Consequently, they have looked for ways to ensure continuity in species protection and safeguards to make sure the achievements so far remain in place. Agreements between governments are regarded as a way of ensuring continuity as they remain in place even if governments, funders or individual people in key positions change. Moreover, they can help in sourcing funding and strengthen the work of NGOs working on those topics. In addition, the CMS offered good possibilities for the participation of professional conservation NGOs and therefore also helps to establish access to governments. This is an important aspect to understand why it was favourable for regional conservation NGOs to work with this particular convention. Second, the CMS Secretariat itself turned its attention to the Southern Cone and promoted the convention with governments. The CMS thus became a framework for regional environmental cooperation as a result of several coinciding factors.

In fact, in several cases, agreements between government agencies already existed previously. Prior to signing the memoranda of understanding under the umbrella of the CMS for both the ruddy-headed goose and the huemul, Argentina and Chile had already signed a treaty on the environment in 1991. However, the objectives were very broad and included amongst other things coordinated action for the protection, preservation, conservation and restoration of the environment as well as the commitment not to carry out unilateral actions that could cause damage to the environment in the other country. The means in order to achieve this include exchange of information regarding legislation and institutions for the protection of the environment, organisation of seminars and bilateral meetings of scientists and experts. In 2002, this was further strengthened with an additional protocol on the conservation of the wild flora and fauna shared between the two countries (Ministerio de Relaciones Exteriores Comercio Internacional y Culto de la Republica Argentina 1991, 2002). While neither the treaty nor the additional protocol refers to any species in particular, the additional protocol does make reference to the CMS and states as one objective the development of memoranda of understanding in the framework of the CMS. Similarly, in the case of the protection of the Andean flamingos, agreements between technical institutions of the different countries already existed before the memorandum under the CMS framework (CMS 2005b, 2008, 14). While these pre-existing agreements are examples of formal cooperation, they were either not very specific or at a lower political level.

Consequently, interviewees clearly saw the CMS as an additional tool to strengthen their objectives, in particular by providing a stronger framework for states to keep their commitments as well as increasing credibility which in turn helps to attract external funding. Interviewees from both, the government and the NGO sector working on the ruddy-headed goose, for example, stated that the added value of the memorandum under the CMS umbrella consists, on the one hand, of providing continuity and a better enforcement of implementation as states would feel more committed if they are accountable to an international organisation, and on the other hand, it helps in providing funding or advice on how to get funding for projects carried out by NGOs.[5] Several interviewees working on the grassland birds memorandum saw the fact that there is a signed commitment by all the states as a major achievement that strengthens their work and given the subsequent development of a formal action plan they were positive that it would be more than just a commitment on paper.[6] Having an agreement signed by the governments also helps promoting the regional work of the NGOs and these frequently refer to the memorandum.[7] One NGO representative also noted that it is easier to get funding from external bodies if something has been signed and potential donors can see that the topic is a priority of the state and not just an NGO:

So once a memorandum exists, something that is signed by the governments...it helps a lot [to get funding]. Because it's a national priority.[8]

Furthermore, the CMS also helps to coordinate and provides some monitoring tools by asking for regular country reports which are publicly available
on its website. In addition, governments that sign a memorandum have to
outline which actions they will take, for example particular conservation
measures or further research, and develop an action plan accordingly. Such
public international commitments thus open up the possibility of "accountability politics" (Keck and Sikkink 1998, 24) by holding governments
to account and reminding them of their commitments.

At the same time, NGOs also use a variety of strategies to pursue their
objectives. Of course, many NGOs do not necessarily work only on
endangered migratory species, but are interested in conservation more
generally. Consequently, they have also worked with other global environmental conventions, such as the Ramsar Convention on Wetlands.
Several of the main sites where the flamingos come together have been
declared Ramsar sites and the regional network therefore also works a lot
with this convention.[9] This means the activities of the different networks
are more or less linked to the CMS. This suggests that the networks
working on species conservation in the region have adopted a rather
pragmatic approach looking for those opportunities and institutional
frameworks that are most likely to strengthen their cause. Nevertheless,
although some initiatives are also linked to other global environmental
conventions, for the Southern Cone region as a whole cooperation has
become most robust in the case of the CMS. This is due to two factors.
First, the CMS institutions offer good possibilities for the participation of
professional conservation NGOs and these also help to gain access to
national governments. Second, the CMS itself has shown more interest in
the Southern Cone region over the last decade.

The CMS strengthens the position of professional conservation NGOs
through its institutional set-up. The convention thus explicitly states that
international and national non-governmental organisations which are
deemed as technically qualified in relation to the conservation of migratory
species may participate as observers unless one-third of the member states
object (CMS 1979, Article VII, 9.). In addition, the CMS also includes a
Scientific Council which was established to provide scientific advice and
make recommendations regarding research activities as well as conservation

and management measures, and in relation to which species should be covered by the CMS. Each member country may appoint a qualified expert as a member of the Scientific Council (CMS 1979, Article VIII). The scientific councillors appointed by the member states can also be experts that do not directly work for the government, but are for example members of NGOs. In the three memoranda where NGOs have played a prominent role, this has been the case for at least one of the countries involved.

In the case of the ruddy-headed goose, the scientific councillor for Argentina was for a long time the director of the *Wetlands International* Office in Buenos Aires.[10] In Bolivia, the link between the GCFA working on the Andean flamingos and the CMS is particularly strong as the scientific councillor appointed by Bolivia is a member of the group. Finally, in Paraguay a member of *Guyra Paraguay*, the Paraguayan branch of *BirdLife International*, was nominated as the Scientific Councillor for the CMS by the government and as one interviewee explained that was why the initiative for the memorandum on grassland birds started in Paraguay:

> Because here in Paraguay, our NGO, which is BirdLife, was better connected with the government. At this moment I was nominated as scientific councillor. So I could go to the level where the government takes decisions and because of this it was Paraguay and it was easier to do the lobbying here than to convince people in other countries.[11]

In this case, the link that the CMS provides between national governments and NGOs was crucial in the development of the memorandum on grassland birds. The scientific councillor from *Guyra Paraguay* whom the government nominated in 2005, had a good relationship with the official in the Paraguayan government dealing with the CMS. Using this opportunity the NGO started to promote the development of a new CMS memorandum relating to a group of grassland birds. These species were already protected under the CMS and regional networks had already carried out various activities for their conservation. Following the lobbying by the NGO and the scientific councillor, the Paraguayan government took up the issue and proposed the development of a new memorandum which was signed in 2007. After the signing of the memorandum, the NGO continued its activities to keep the memorandum alive and to promote the development of an action plan. This included for example organising meetings, identifying potential sources of funding and keeping partner NGOs in the region informed.[12]

While regional conservation networks have been fundamental in working on conservation and awareness raising as well as cooperating with governments, encouragement from the CMS Secretariat itself was also crucial in further promoting the convention with national governments and establishing the CMS as a framework for regional cooperation. In relation to this, two developments are important. First, in the last decade the CMS has undergone some internal changes that are not directly related to the Southern Cone, but which have had an impact on the region. While in the early 1990s commentators pointed out that a decade after the Convention entered into force, only a few agreements were signed (de Klemm 1994, 71), 10 years later a drive towards the further implementation of the convention and the conclusion of new agreements has been noted. However, Caddell (2005, 140) also points to the Eurocentricity of the convention as most of the agreements signed by 2005 still related to species that passed through European countries on their migratory routes. Nevertheless, the objective of the CMS Secretariat is to attain a global coverage and the number of agreements as well as the number of countries joining the CMS has increased even more from 2006 onwards (CMS 2007, 2; Lee et al. 2011, 18, 24). The sudden increase of memoranda signed between Southern Cone countries from 2006 onwards thus also has to be seen in the context of a general trend where the CMS Secretariat promoted the convention beyond Europe and Africa where most of its activities had been taking place initially. At the same time cooperation between the CMS and Southern Cone governments was helped significantly by a second aspect.

A member of staff of the convention's secretariat came from the Southern Cone and had been working in the region previously. A biologist by training, the official had been working on migratory species and in particular grassland birds already before working for the CMS and was therefore already part of the network of researchers that had developed in the region on this issue. Moreover, being a native Spanish speaker made it easier to interact with governments in the region. The official put a lot of effort into contacting the governments to try and convince them to come to meetings and make them aware of the funding possibilities through the CMS.[13] With an official who knew the Southern Cone context well, the CMS Secretariat was in a much stronger position to approach the Southern Cone governments and promote the convention. Overall, this confirms findings of other studies on the autonomous influence of international bureaucracies (Biermann and Siebenhüner 2009a). The CMS Secretariat as a whole has had an important influence on agenda-setting by actively promoting the convention outside Europe. However, in addition the individual people working for a bureaucracy matter as well and this helps to explain why the CMS was particularly successful in targeting Southern Cone governments. This case study therefore also demonstrates how international bureaucracies can contribute to shaping the characteristics of regional environmental cooperation by promoting a particular institutional framework or environmental concern and turning their attention to certain governments or regions.

However, it is more likely for international bureaucracies to develop an independent influence if a particular environmental concern is not perceived as very urgent or salient by governments and in particular the most powerful governments (Biermann and Siebenhüner 2009b, 335). This is clearly the case for migratory species in the Southern Cone and stands in stark contrast with the issue of water which is a much more prominent and sensitive issue in the region as presented in the previous chapter. This is arguably one of the reasons why regional conservation networks together with the CMS Secretariat were able to exert significant influence and promote the protection of migratory species with governments in the region. In both, the case study on the La Plata basin regime and in Mercosur civil society organisations have been much less successful in linking their initiatives to formal cooperation by governments and there is a division between regional civil society initiatives on the one hand and cooperation involving governments on the other. The relatively strong links between governments, NGOs and the CMS itself are a key reason why this case study represents a relatively robust example of regional environmental cooperation in the Southern Cone, although this is also possible only under very specific conditions. The final section examines in more detail the elements that make cooperation more robust in this case, but also outlines some important limitations.

Sources of Strength and Limitations

The relatively close links and cooperation between endogenous and a variety of exogenous drivers are one of the main sources of strength of this case study. This is very different from the previous case study which was characterised by the sidelining of interested civil society networks, a strong dependence on one particular funder as well as public distrust of that funder. Nevertheless, important limitations also remain in this case study. In particular, all the memoranda of understanding are non-binding and implementation is often challenging and suffers from a lack of funding and government commitment.

The relatively close links between the different endogenous and exogenous drivers promoting regional environmental cooperation in this case study are to a large extent due to the existence of relatively dense regional conservation networks committed to species protection with members working for NGOs, universities, national park administrations and environmental agencies of national governments as well as the CMS Secretariat. Moreover, in some cases, individual members of a network have also moved between different positions. A government official working for the Argentinean Environment Secretariat and dealing with the CMS amongst other things was, for example, a biologist by training and had previously worked for an NGO.[14] In another case, an ornithologist had previously worked in Uruguay and then moved on to work for the CMS Secretariat in Bonn.[15] Finally, the scientific councillor of Paraguay, equally a biologist by training who previously worked for the NGO *Guyra Paraguay*, was appointed Minister of Environment in August 2013 (CMS 2013). Moreover, researchers are often also part of international communities in their area of expertise and participate in international conferences, exchanges of information and so on and are therefore in a strong position to access international sources of funding and expertise.

Networking is further facilitated by the CMS itself. The fact that governments can appoint NGO members as scientific councillors strengthens the link between NGOs and governments institutionally going beyond personal connections. This set-up provides advantages for both sides. NGOs and networks of researchers gain a position that is very close to policy-makers and are able to bring in their ideas and remain updated for the latest developments with regard to policy-making. Even if only one country designates a scientific councillor from a particular network, information can then be quickly shared among the whole network. On the other hand, the position of scientific councillor is voluntary and unpaid.[16] This means governments which may not have a lot of resources to dedicate to species conservation can benefit from the expertise and resources of a particular NGO and the personal commitment and the networks of individual researchers and this helps governments keep international obligations or demonstrate their commitment to an environmental concern.

Dense networks are thus one of the key reasons why regional environmental cooperation is the most robust in this case study. Moreover, conservation networks have been successful in securing external funding for their activities. It is notable that funding comes from a variety of different sources, including international conventions such as the CMS and Ramsar, development banks, international NGOs and foundations as well as Northern government agencies and private companies. While there are some larger projects, in many other cases funding consists of relatively small amounts. Consequently, the influence of any single funder on the characteristics of cooperation is much more limited than in the cases of the La Plata basin regime and Mercosur, although donor priorities have also had some influence. Interviewees have noted, for example, that it is easier to get funding for species which migrate between North and South America as there is more funding available for these from US and Canadian bodies. As a result, these species are also much better researched and understood than species that migrate within South America only.[17] Another example is the case of the grant that the Argentinean NGO *Aves Argentinas* received from the GEF. While the original project proposals included the south of Brazil, Uruguay and parts of Paraguay, this was rejected twice. At the third attempt, the NGO received significant feedback from the World Bank, including the recommendation to apply for one country only and finally received the funding for Argentina which has the biggest share of the ecosystem concerned. The World Bank thus did not support a regional approach, which from an environmental point of view would have made more sense, apparently because this would have been more complicated in terms of administration.[18] These examples demonstrate donor preferences on individual issues, but the overall influence of any one funder on the characteristics of cooperation remains relatively low compared to the La Plata basin regime and Mercosur. In addition, because funding comes from a variety of different sources, it is also less dependent on any one single funder and therefore less vulnerable should funding from that source stop.

Furthermore, regional conservation networks have also used a variety of tools and strategies going beyond the CMS framework, and this further contributes to a higher level of robustness. This is particularly well developed in relation to the protection of grassland birds and their habitats. In this case, the species' habitat is largely in private areas, so that here the NGOs involved have used different strategies and have looked for ways of working with landowners. According to some reports less than 1% of the grasslands in the Southern Cone are protected areas (BBC 2011) and the vast majority are in private hands. The NGOs therefore quickly concluded that the only way to preserve the grasslands was to continue agricultural production and to work with the producers:

> So it was clear that we had to do conservation while continuing [agricultural] production. So the Grasslands Alliance initiative was created where we're looking for allies in the productive, academic and governmental sector. We think that the only way to conserve these areas is to continue production.[19]

Consequently, as an alternative, the NGOs in the region have looked for a way to continue using the land for agricultural purposes while at the same time preserving it and protecting the species that live there. Instead of using the land for monocultures or intensive cattle ranching, the NGOs are developing models for raising cattle on the natural grasslands in a sustainable way. In order to make this idea attractive for landowners and farmers they have established links with the governments of Uruguay and Paraguay as well as the governments of several provinces and federal states in Argentina and Brazil with the objective of creating economic incentives to support this way of cattle ranching. These include financial or tax incentives for farms which are managed in such a way that they provide ecosystem services, but also the development of a certification scheme for beef produced in natural grasslands in order to increase the market value of the product. To give more structure to this work at a regional level the partner organisations of *BirdLife International* in Argentina, Brazil, Paraguay and Uruguay have supported the creation of the *Alianza del Pastizal* (Alliance for the Grasslands) with the objective of monitoring grassland birds and working with rural producers in order to achieve more sustainable production. The alliance builds on the idea that more sustainable production is possible if corresponding models and policies exist. Moreover, it draws strength from the fact that many farmers have reluctantly turned to producing soy as this provides a better income, but in fact prefer raising cattle as their families have done for generations. Some rural producers are thus keen on the prospect of finding an alternative, economically viable, way of returning to cattle ranching. The *Alianza del Pastizal* works with a range of experts and universities in the region and holds regular meetings with producers and other stakeholders. Regular updates are made available on its website. In order to support these activities and to carry out pilot projects, the NGOs have successfully applied for funding from different bodies, such as the GEF or the IDB.[20]

In addition, regional conservation networks have in some cases also adopted a pragmatic approach and cooperated with large transnational companies whose activities are also a main threat to the species concerned. The group of conservationists working on the Andean flamingos, has, for example, received funding from Rio Tinto, a mining company operating in the Andes. *BirdLife International* had already established a programme with Rio Tinto and they approached the GCFA in order to monitor the flamingos and find ways of mitigating the negative impact of mining on the birds (BirdLife International 2008; Marconi 2010, 38). Similarly, *Aves Uruguay*, the Uruguayan partner of *BirdLife International* approached the Finnish company UPM, one of the largest companies producing pulp for the paper industry in Uruguay, to discuss possible cooperation. While *Aves Uruguay* is generally not in favour of the paper industry because of the monocultures with many environmental impacts, a representative of the NGO noted that it would be difficult to change the paper industry in Uruguay:

> ...we realise that a few NGOs or the environmental sector will not change the policy in this way. We don't have the strength and the capacity to do this. ...So we're looking for opportunities in this sector.[21]

Aware of the company's need to promote a greener image, *Aves Uruguay* thus started discussions with one of the companies in order to discuss whether it is possible to use the land that is not suitable for plantations for conservation. Finally, *Guyra Paraguay* benefited from funding of the binational power company Yacyretá that constructed a large hydropower dam on the border between Paraguay and Argentina which resulted in the flooding of a large area discussed in the previous chapter.[22]

Conflicts of interests between intensive resource exploitation and conservation are clearly evident also in this case study, but the NGOs and networks involved have generally taken a pragmatic approach and attempted to work with governments and companies in order mitigate some of the environmental damage or carve out areas for conservation while open and public criticism of governments is rare. One interviewee made this particularly clear by stressing that the approach of *Aves Argentinas* has historically always been to work with the government and to make recommendations and give advice this way rather than openly criticising the government and that this constituted an important difference in comparison with other environmental organisations:

And so nowadays we see ourselves not as an entity that denounces, criticises and very quickly gives an opinion regarding the problems that governments face, but instead we see ourselves as partners, collaborators in the development of the policies of the state which in fact address the common good. And that's the way we position ourselves in the political arena. With the vision of cooperation with the governments and not confrontation.

At the same time the interviewee also noted that this can blur the boundary between the work of NGOs and the work of the government:

We don't formally work with the governments, but we are an entity that is frequently consulted...I think... because of a series of crises that the countries suffered in the last 30 or 40 years...that is why NGOs flourished so much in this part of the world. In a way the NGOs occupy the space that public offices don't occupy.[23]

Another interviewee explained:

But in fact we are facilitating the work [of governments], to comply with a mission that they have, a commitment of the government that they have towards the CMS. What I think is that if there is no support from the NGOs as it is now, it will break down. It's possible that the interest continues, the governments are interested, that's why they signed, but they don't give it priority.[24]

The success of NGOs in promoting regional environmental cooperation has thus been dependent on a willingness to work with governments and within the existing institutions as well as the significant scientific and financial resources of the NGOs. They have access to these resources in part due to their links to international networks, but in part also due to working with and accepting funding from large companies. Although several of the initiatives, most notably the *Alianza del Pastizal,* in some respects work towards constructing alternative models of natural resource governance, the issues tend to be framed in technical and scientific terms and remain within dominant paradigms. The *Alianza del Pastizal* thus frequently refers to the value of ecosystem services which natural grasslands provide. This fits with the "green economy" approach which remains within the framework of capitalist market-based structures and which tends to be dominant among policy-makers in Latin American governments (de Castro et al. 2016, 9–10). This is a noticeable difference in particular in comparison with the regional civil society networks working on issues related to water presented in the previous case study, which built their arguments around water and land rights. These often reflect broader environmental justice discourses combining environmental and social issues which also raise critical questions regarding the development model adopted by governments. The emphasis on technical expertise is also shared by the CMS. Participation in meetings is thus not open to any civil society organisation with an interest in the issue. Instead, the convention clearly specifies that organisations need to be "technically qualified in protection, conservation and management of migratory species" and approved by the country in which they are located (CMS 1979, Article VII, 9.). This stipulation also appears to give national governments some scope to exclude organisations which they do not wish to gain any influence.

Furthermore, even though there have been important increases in cooperation under the CMS umbrella over the last decade, there are clear indications that the protection of migratory species remains a low political priority in the Southern Cone region. In particular, the four memoranda that have been signed are all non-binding and do not entail financial commitments to the CMS.[25] Governments have thus not made the step towards establishing legally binding agreements with their own institutional structures and financial contributions as is the case for CMS agreements in other parts of the world. In addition, interviewees have also pointed to other significant challenges which still remain. Interviewees working on the protection of the Andean flamingos, for example, have pointed out that it is very difficult to establish new protected areas and national as well as provincial authorities constantly need to be reminded of existing protected areas in order to avoid the development of economic activities in these areas. The conservation of the flamingos is therefore very much dependent on the work of the regional conservation network and leading figures within the group and would not be guaranteed by national or provincial governments on their own, even if they have on paper made relevant commitments. According to one interviewee:

> It's not the countries themselves, this didn't come up from policies of the four countries. Instead it developed from the researchers who got together and from that we're trying to push for policies in the different countries and provinces.[26]

Another interviewee explained:

> For me the initiative works because in fact the drivers of the initiative are several people who, independently from the institutions where they are, are interested in a conservation strategy for flamingos. ... If you only left it to the institutions I think it wouldn't work.[27]

Similarly, in relation to the grassland birds memorandum one NGO representative stated that the governments still find it difficult to implement the action plan by themselves and if there is not constant support from another actor or organisation things would not move a lot. Another issue is the constant lack of funding, which means that even with the action plan agreed there is no money from the government to implement it and NGOs or other actors have to look for funding for the different activities.[28]

Finally, some range states have decided not to join a particular memorandum or there were delays. The memorandum on the Andean flamingos has only been signed by Bolivia, Chile and Peru. Argentina is a range state, but has so far chosen not to sign, although Argentinean researchers and national park administrations regularly participate in research and conservation activities. Similarly, Bolivia participated in the grassland birds memorandum from the beginning, but only signed it two years later in 2009. Moreover, Brazil as the largest and most powerful country in the region only became a party to the CMS relatively late in 2015, although it had signed the memorandum on grassland birds and actively participated in its development prior to this. The precise reasons of why some countries have abstained from making formal commitments, although some of their environmental agencies do engage in cooperation in practice, have been difficult to establish. Interviewees have mostly pointed to rather general internal administrative or coordination difficulties, rather than specific obstacles or disagreements. However, it is important to note that the protection of migratory species and their habitats often conflicts with important economic activities and this is likely to play a role in preventing or delaying formal commitments on the part of governments. For example the habitat of Andean flamingos often corresponds to areas that are important for mining while one of the main threats to the grassland birds is intensive agriculture and in particular soybean monocultures as well as pine and eucalyptus plantations for the paper industry (BBC 2011; Caziani et al. 2007; CMS 2012, 4). One of the threats to the huemul on the other hand is the fragmentation of its habitat due to large infrastructure developments such as hydropower installations (The Guardian 2011). Tensions between neo-extractivist development strategies and environmental conservation are therefore evident also in this case study.

NOTES

1. Interview, Asociación Guyra Paraguay, Asunción, 2011.
2. Interview, Universidad Nacional de Salta, Salta, 2011, see also GCFA (2011), Marconi (2010, 37).
3. Interview, Universidad Nacional de Salta, Salta, 2011, author's translation.
4. Marconi (2010, 37–38); interviews, Universidad Nacional de Salta, Salta, 2011; Administración de Parques Nacionales, Salta, 2011.
5. Interviews Wetlands International, Buenos Aires, 2011; Secretaría de Ambiente y Desarrollo Sustentable, Buenos Aires, 2011.
6. Interviews, Secretaría de Ambiente y Desarrollo Sustentable, Buenos Aires, 2011; Secretaría del Ambiente, Asunción, 2011; Asociación Guyra Paraguay, Asunción, 2011.
7. Interview, Aves Argentinas—A.O.P., Buenos Aires, 2011.
8. Interview, Asociación Guyra Paraguay, Asunción, 2011, author's translation.
9. Interview, Universidad Nacional de Salta, Salta, 2011.
10. Interview, Wetlands International, Buenos Aires, 2011.
11. Interview, Asociación Guyra Paraguay, Asunción, 2011, author's translation.
12. Interviews, Asociación Guyra Paraguay, Asunción, 2011; Aves Argentinas—A.O.P, Buenos Aires, 2011.
13. Interview, CMS Secretariat, Bonn, 2011.
14. Interview, Secretaría de Ambiente y Desarrollo Sustentable, Buenos Aires, 2011.
15. Interview, CMS Secretariat, Bonn, 2011.
16. Interview, Asociación Guyra Paraguay, Asunción, 2011.
17. Interviews, Asociación Guyra Paraguay, Asunción, 2011; Wetlands International, Buenos Aires, 2011; see also Di Pangracio, Rabufetti, and Grilli 2011: 494.
18. Interview, Aves Argentinas—A.O.P., Buenos Aires, 2011.
19. Interview, Aves Uruguay, Montevideo, 2011, author's translation.
20. Interviews, Aves Argentinas—A.O.P., Buenos Aires, 2011; Aves Uruguay, Montevideo, 2011; Secretaría de Ambiente y Desarrollo Sustentable, Buenos Aires, 2011.
21. Interview, Aves Uruguay, Montevideo, 2011, author's translation.
22. Interview, Asociación Guyra Paraguay, Asunción, 2011.

23. Interview, Aves Argentinas—A.O.P., Buenos Aires, 2011, author's translation.
24. Interview, Asociación Guyra Paraguay, Asunción, 2011, author's translation.
25. Interview, CMS Secretariat, Bonn, 2011.
26. Interview, Universidad Nacional de Salta, Salta, 2011, author's translation.
27. Interview, Administración de Parques Nacionales, Salta, 2011, author's translation.
28. Interview, Asociación Guyra Paraguay, Asunción, 2011.

References

Balsiger, Jörg, Miriam Prys, and Niko Steinhoff. 2012. *The Nature and Role of Regional Agreements in International Environmental Politics: Mapping Agreements, Outlining Future Research*. GIGA Working Paper 208. Hamburg: German Institute of Global and Area Studies. http://www.giga-hamburg.de/index.php?file=workingpapers.html&folder=publikationen.

Bauer, Steffen. 2009. The Desertification Secretariat: A Castle Made of Sand. In *Managers of Global Change—The Influence of International Environmental Bureaucracies*, ed. Frank Biermann and Bernd Siebenhüner. London: MIT Press.

BBC. 2011. El Cono Sur Se Queda Sin Pastizales Y Sin Aves Migratorias, 12 January 2011. http://www.bbc.co.uk/mundo/noticias/2011/01/110112_paraguay_aves_migratorias_peligran_vh.shtml.

Biermann, Frank, and Bernd Siebenhüner (eds.). 2009a. *Managers of Global Change—The Influence of International Environmental Bureaucracies*. London: MIT Press.

Biermann, Frank, and Bernd Siebenhüner. 2009b. The Influence of International Bureaucracies in World Politics: Findings from the MANUS Research Program. In *Managers of Global Change—The Influence of International Environmental Bureaucracies*, ed. Frank Biermann and Bernd Siebenhüner. London: MIT Press.

BirdLife International. 2008. Working Together to Conserve Flamingos at a Network of Sites in the High Andes. http://www.birdlife.org/datazone/sowb/casestudy/292.

Caddell, Richard. 2005. International Law and the Protection of Migratory Wildlife: An Appraisal of Twenty-Five Years of the Bonn Convention. *Colorado Journal of International Environmental Law and Policy* 16: 56–113.

Caziani, Sandra M., Omar Rocha Olivio, Eduardo Rodríguez Ramírez, Marcelo Romano, Enrique J. Derlindati, Andrés Talamo, David Ricalde, Carmen Quiroga, Juan Pablo Contreras, Mariana Valqui and Heber Sosa. 2007.

Seasonal Distribution, Abundance, and Nesting of Puna, Andean and Chilean Flamingos. *The Condor* (109): 87–276.

CMS. 1979. Convention on the Conservation of Migratory Species of Wild Animals.

CMS. 2002. *National Report Argentina 2002.* http://www.cms.int/documents/national_reports/index_by_cop.htm.

CMS. 2005a. *National Report Argentina 2005.* http://www.cms.int/documents/national_reports/index_by_cop.htm.

CMS. 2005b. *Reunión Regional América Latina Y El Caribe.* http://www.cms.int/bodies/regional_meetings_mainpage.htm.

CMS. 2006. CMS Accession Guidelines January 2006—How to Become a Member of CMS and CMS-Related Agreements. www.cms.int.

CMS. 2007. *32nd Meeting of the Standing Committee—Towards the Future Shape of CMS.* http://www.cms.int/bodies/future_shape/future_shape_mainpage.htm.

CMS. 2008. *National Report Peru 2008.* http://www.cms.int/documents/national_reports/index_by_cop.htm.

CMS. 2012. *CMS Bulletin 5/6 2012.* http://www.cms.int/publications/CMS_Bulletin_2012.html.

CMS. 2013. Cristina Morales Consejera Científica de CMS Fue Nombrada Ministra Del Ambiente Del Paraguay. http://www.cms.int/news/PRESS/nwPR2013/08_aug/nw_300813_cristina_morales.html.

CMS. 2016. Parties and Range States. http://www.cms.int/en/parties-range-states.

Corporación Nacional Forestal. 2010. Junto a La Comunidad, CONAF Realiza Censo de Huemules En RN Tamango, 13 April 2010. http://www.conaf.cl/buscar.html?keyword=huemul.

Corporación Nacional Forestal. 2011a. CONAF Y Carabineros Unidos En La Protección Del Huemul, 2 February 2011. http://www.conaf.cl/buscar.html?keyword=huemul.

Corporación Nacional Forestal. 2011b. Comunidad Pedalea En Defensa Del Huemul, 6 April 2011. http://www.conaf.cl/buscar.html?keyword=huemul.

Comité Nacional Pro Defensa de la Flora y Fauna Chile. 2013. CODEFF Chile. http://www.codeff.cl.

De Castro, Fabio, Barbara Hogenboom, and Michiel Baud. 2016. Introduction: Environment and Society in Contemporary Latin America. In *Environmental Governance in Latin America*, ed. Fabio de Castro, Barbara Hogenboom, and Michiel Baud. Basingstoke: Palgrave Macmillan.

De Klemm, Cyrille. 1994. The Problem of Migratory Species in International Law. In *Green Globe Yearbook of International Co-Operation on Environment and Development 1994*, ed. Helge Ole Bergesen and Georg Parmann. Oxford: Oxford University Press.

GCFA. 2011. Grupo Conservación Flamencos Altoandinos. www.flamencosandi-nos.org.

Guardian. 2011. Chile Approves $7bn Hydroelectric Dam in Patagonian Wilderness. 10 May 2011.

Huilo Huilo Foundation. 2013. Huilo Huilo Foundation. http://www.huilohuilo.com/en/fundacion/conservation-center-south-huemul.

Keck, Margaret E., and Kathryn Sikkink. 1998. *Activists Beyond Borders—Advocacy Networks in International Politics*. London: Cornell University Press.

Lee, Robert, Begonia Filgueira, and Lori Frater. 2011. *Convention on Migratory Species: Future Shape Phase III*. Southampton. http://www.cms.int/bodies/future_shape/future_shape_mainpage.htm.

Marconi, Patricia. 2010. *Manual de Técnicas de Monitoreo de Condiciones Ecológicas Para El Manejo Integrado de La Red de Humedales de Importancia Para La Conservación de Flamencos Altoandinos*. Salta: Fundación Yuchan. http://redflamencos.blogspot.com.ar/.

Ministerio de Relaciones Exteriores Comercio Internacional y Culto de la Republica Argentina. 1991. *Tratado Entre La Republica Argentina Y La Republica de Chile Sobre Medio Ambiente*. http://tratados.cancilleria.gob.ar/.

Ministerio de Relaciones Exteriores Comercio Internacional y Culto de la Republica Argentina. 2002. *Protocolo Específico Adicional Sobre Conservación de La Flora Y Fauna Silvestre Compartida Entre La Republica Argentina Y La Republica de Chile*. http://tratados.cancilleria.gob.ar/.

Najam, Adil. 2004. Dynamics of the Southern Collective: Developing Countries in Desertification Negotiations. *Global Environmental Politics* 4 (3): 54–128.

Secretaría de Ambiente y Desarrollo Sustentable de la Nación—Argentina. 2002. *Plan Nacional de Conservación Y Recuperación Del Huemul (Hippocamelus Bisulcus) En Argentina*. http://www.ambiente.gov.ar/archivos/web/Phuemul/File/Plan_nacional_huemul.pdf.

Secretaría de Ambiente y Desarrollo Sustentable de la Nación—Argentina. 2008. *Plan Nacional de Conservación Y Recuperación Del Huemul En Argentina-Acta de Reunión – Marzo 2008*. http://www.ambiente.gov.ar.

Secretaría de Ambiente y Desarrollo Sustentable de la Nación—Argentina. 2009. *Plan Nacional de Conservación Y Recuperación Del Huemul En Argentina-Acta de Reunión – Junio 2009*. http://www.ambiente.gov.ar.

Selin, Henrik. 2012. Global Environmental Governance and Regional Centers. *Global Environmental Politics* 12 (3): 18–37.

Serret, Alejandro. 2001. *El Huemul – Fantasma de La Patagonia*. Ushuaia: Zagier & Urruty Publications.

Steinberg, Paul F. 2001. *Environmental Leadership in Developing Countries—Transnational Relations and Biodiversity Policy in Costa Rica and Bolivia*. London: MIT Press.

Steinberg, Paul F. 2003. Understanding Policy Change in Developing Countries: The Spheres of Influence Framework. *Global Environmental Politics* 3 (1): 11–32.

Wetlands International. 1998. *Desarollo de Un Plan de Conservación Para El Cauquén Cabeza Colorada (Chloephaga Rubidiceps), En La Región Austral de Argentina Y Chile – Informe Final.*

Wetlands International. 2000. *Medidas de Accion Para La Conservación Del Cauquén Colorado (Chloephaga Rubidiceps), En Argentina Y Chile – Informe Final Para La CMS.*

Wetlands International. 2004. *Concerted Actions for the Management and Conservation of the Ruddy-Headed Goose (Chloephaga Rubidiceps) in Chile and Argentina—Final Report.*

Wetlands International. 2009. *Ruddy-Headed Geese (Chloephaga Rubidiceps) in Danger: Population Status and Conservation Actions in Argentina and Chile—Final Report.* http://www.cms.int/publications/pdf/Cauquen_colorado_june2009_en.pdf.

Vila, Alejandro R., Rodrigo López, Hernán Pastore, Ricardo Faúndez, and Alejandro Serret. 2006. Current Distribution and Conservation of the Huemul (hippocamelius Bisulcus) in Argentina and Chile. *Mastozoología Neotropical* 13 (2): 263–269.

CHAPTER 6

Conclusion

The Southern Cone of South America presents a fascinating puzzle for the study of regional environmental cooperation due to the combination of two aspects. On the one hand, regional environmental cooperation has increased in quantity and quality since the early 1990s. Yet, on the other hand, it has a relatively low profile and takes place in the margins of other cooperation efforts and political priorities. In order to understand the development and nature of regional environmental cooperation in the Southern Cone, the book started out by seeking to examine three research questions: First, what does regional environmental cooperation in the Southern Cone consist of? Second, who has promoted it over the last two decades and how, or what has been the process leading to regional environmental cooperation? And finally, what are important characteristics? Based on the analysis of these questions I have outlined a broad pattern identifying components of cooperation and the process leading to robust cases of cooperation. Reflecting findings from other studies on regions in the South, an important characteristic of environmental cooperation in the Southern Cone is its marginality. This is evident in three key elements; the absence of regimes created specifically in order to address a particular regional environmental concern; a high dependence on external funding; and the vague and non-binding nature of agreements between countries. However, the book argues that the marginality of regional environmental cooperation can be overcome to some extent if the regular activities that make up cooperation in practice are linked with formal commitments by states. This makes regional environmental cooperation more robust and therefore more likely to continue. The process

© The Author(s) 2017   151
K.M. Siegel, *Regional Environmental Cooperation in South America*,
International Political Economy Series, DOI 10.1057/978-1-137-55874-9_6

leading to robust cases of regional environmental cooperation linking formal agreements with regular activities has been shaped by various actors at different stages. Initially, a demand for regional environmental cooperation emerged from an increased interest and knowledge of the transboundary dimensions of an environmental concern. Networks of researchers and civil society organisations are an important endogenous driver at this stage. However, as governments only make limited funding available, exogenous drivers in the form of external funders become a crucial actor in making cooperation in practice possible. As a result, cooperation is shaped to varying degrees by the objectives and priorities of external funders. In addition, governmental approval is necessary for written agreements between states, and governments decide which topics, approaches and framings proposed by endogenous or exogenous drivers are taken up in formal cooperation.

In addition to establishing this broad pattern of regional environmental cooperation in the margins, a comparison of the two case studies and the regional organisation Mercosur which initially seemed a promising framework for environmental cooperation, also uncovers several noteworthy observations about the process and nature of environmental cooperation in the Southern Cone which point to important questions for further research in South America and beyond. This final chapter focuses on three key findings that have emerged from the analysis; first, the findings from the book show that while there are various different endogenous and exogenous drivers seeking to promote regional environmental cooperation, there are considerable asymmetries in their access and influence in being able to shape the processes leading to cooperation. In particular, external funders play a central, but often contested role which has led to mixed outcomes. Moreover, there are significant differences between domestic civil society organisations and the groups that have been able to exert most influence have been those focussing on technical and scientific framings of environmental problems and able to deploy scientific, human and financial resources to support governments. These findings are perhaps not surprising, but they are important because they shape the nature of cooperation and can also have implications for the perceived legitimacy of cooperation. Second, the findings from the book show that government-led regional integration processes are not necessarily a driver for regional environmental cooperation and on the contrary, they can also present obstacles. This is somewhat more unexpected as it is the opposite of what has been observed elsewhere, certainly in the case of the EU, but also in other regions where processes of regional integration have sometimes led to increased environmental cooperation (Elliott and Breslin 2011, 15; Siegel 2016a, 719–720). Finally, the region's position in the global political economy as an exporter of primary commodities to other parts of the world acts as a significant constraint on environmental governance because intensive natural resource exploitation is closely related to socio-environmental problems, but the reliance on resource extraction for economic and social development makes it difficult to address environmental concerns. This is important because it means that it will be difficult for South American governments to strengthen environmental cooperation without the support of those buying and consuming natural resources from South America.

Unequal Influence in the Process Leading to Regional Environmental Cooperation

There is a range of different actors that are interested in addressing shared and transboundary environmental concerns and therefore constitute potential drivers for regional environmental cooperation in South America. This includes endogenous drivers, that is actors from within the region, such as civil society organisations ranging from grassroots movements to professional NGOs as well as networks of researchers and staff working for government agencies. Moreover, there are also exogenous drivers, meaning actors from outside the region, such as various donors and international organisations which seek to promote regional environmental cooperation for a variety of reasons. However, the findings from the book demonstrate that the influence of these different drivers in the process leading to regional environmental cooperation differs significantly. Some drivers therefore succeed in convincing governments that their concerns should be addressed in formal cooperation and through regular activities and work with governments towards this. For others, access to governments and international and regional institutions is much more limited, and consequently, their perspectives are reflected to a much lesser extent in regional environmental cooperation, if at all. The cases examined in the book demonstrate two trends in particular. First, due to the reliance on external funding in all cases, external donors hold a privileged position to shape regional environmental cooperation according to their priorities and objectives. However, this is not uncontested and disagreements and distrust of foreign funders can also present obstacle or undermine regional environmental cooperation. Second, governments focus on technical and scientific approaches to regional environmental cooperation rather than the often more contentious social or political questions. As a result, civil society organisations focussing on technical aspects that are able to offer governments their scientific expertise are much better placed to shape regional environmental cooperation than those focussing on social and environmental justice. While neither of these findings is unexpected, they are important because they shape regional environmental cooperation in particular ways and also raise important questions regarding its legitimacy.

As government funding is limited external funding plays a prominent role in the process leading to regional environmental cooperation in the Southern Cone as set out in Chap. 1. External funding supports the regular activities that make up cooperation in practice in all cases and in the case of Mercosur and the CMS it is also very evident that exogenous drivers have played an important role in promoting the agreements and joint declarations that make up formal cooperation. It is not surprising then that the objectives and priorities of funders shape the characteristics of cooperation. However, the analysis also shows that in the Southern Cone a range of different funders with very different objectives and approaches are involved in regional environmental cooperation. This has also resulted in varying relationships with the Southern Cone governments and civil society with disagreements and distrust being evident in some cases.

As set out in Chap. 3 in the case of Mercosur European funders have been a crucial source of support for the regional organisation including on environmental topics. The objective of European funders has been to strengthen regional cooperation modelled on the EU. This has been supported by the belief that the EU model is valuable for other regions, but no doubt the projection of the EU model in other parts of the world also serves the purpose of enhancing the EU's reputation globally. European funding has been crucial to support regular meetings and projects on environmental issues between the Mercosur countries and this has also led to some joint declarations. However, Mercosur's environmental agenda has suffered from a lack of continuity which undermines the progress that has been made and friction in the relationship between the Mercosur governments and European funders is one of the aspects accounting for this. It has thus become evident over time that the Mercosur governments do not wish to follow the EU's path of creating supranational institutions and Mercosur's environmental institutions have remained very weak. Moreover, European funders have been more concerned with promoting a particular model of regional integration than with addressing any particular regional environmental problem. Unlike the other two case studies, regional environmental cooperation in the Mercosur framework has not been driven primarily by a desire to address a specific shared environmental concern.

In the case of the La Plata basin examined in Chap. 4 regional environmental cooperation focuses on various aspects relating to sustainable development in the basin. Here, the GEF as the main funder has been crucial in making regional environmental cooperation possible and consequently the priorities of the GEF have shaped cooperation in the basin. One of the main criteria of the GEF is that projects need to be of global importance, and as a result, some parts of the basin that were not deemed to meet this criterion have not received funding. Moreover, several researchers and organisations have reported public distrust of GEF projects. This is due to the GEF's link to the World Bank which has a very negative reputation in the region since the imposition of structural adjustment packages during the neoliberal period. Moreover, water governance is a sensitive topic that has been the subject of political contestation in several contexts in the region. Finally, in regional environmental cooperation on migratory species discussed in Chap. 5 the overall impact of external funders has been least significant as there was no single dominant donor. In this case, the CMS also fulfilled other important functions that promoted cooperation on migratory species in South America. The initiative of the secretariat was thus important in approaching governments to convince them of the value of working in the framework of the convention and the regular report and updates required by the CMS ensure a continuous engagement and publicly available information.

On the whole, the case studies in this book demonstrate that external funders have a significant say in what types of projects get funded, where these are located and what the objectives are. As a result, the influence of actors from outside the region in the processes leading to regional environmental cooperation in the Southern Cone is more pronounced than in environmental cooperation in the global North. It is likely that this is also the case elsewhere as the majority of studies on environmental cooperation in regions of the South have found that exogenous drivers play a central role and some have concluded that external actors are more important in driving regional environmental cooperation than domestic ones (Compagnon et al. 2011, 107; Kulauzov and Antypas 2011, 113). However, the different cases also show that relationships between governments, civil society and external funders can play out in very different ways. The involvement of external funders can therefore also present obstacles and it adds a further level of potential contestation. This echoes findings from previous studies on environmental aid which have highlighted that the role of external funders is often contested because funders are of course not neutral, but usually have strong interests of their own and in many cases these dominate environmental aid programmes and cooperation (Connolly 1996, 329; Fairman and Ross 1996, 42). Foreign funders can also have a considerable impact on domestic politics to the extent that they can affect what types of domestic environmental groups exist and which ones succeed as well as shaping their agendas and strategies (Lewis 2016, 10). Moreover, distrust of foreign funders is not unusual. In her study on environmentalism in Ecuador Lewis for example labels transnational funders as "ecoimperialists" echoing the language used by many Ecuadorians who view transnational funders as foreign intruders imposing agendas in pursuit of their own interests and interfering with domestic policies and approaches to development (Lewis 2016, 48–49). It is not surprising then that the donor-recipient relationship is often complex and the source of many disputes (Connolly 1996; Fairman and Ross 1996). However, there are many different types of funders that might promote regional environmental cooperation including for example UN agencies, international organisations and international development banks as well as national government agencies of Northern countries, NGOs and private foundations. More systematic comparisons to establish what types of funders are active in different regions, what their objectives and relationships to governments and civil society are and how these affect regional environmental cooperation would thus significantly advance our understanding of environmental cooperation in regions where there is a high dependence on external funding.

In addition to exogenous drivers, there is a range of potential endogenous drivers for regional environmental cooperation in South America. However, a comparison of the different cases examined in the book and their evolution over time suggests that there are considerable asymmetries in the influence that different domestic environmental groups have in the process leading to regional environmental cooperation due in part to differences in resources, but also due to the preferences of national governments. As a result, only some groups manage to put their concerns on the agendas of international and regional institutions and promote formal cooperation between governments on those issues. In the case of Mercosur there was initially more interest on the part of environmental groups and these also had some, albeit limited, access to Mercosur's environmental institutions and in particular the SGT6. However, this interest declined over time as it became evident that Mercosur has not been given the mandate to deal with many of the region's most important transboundary environmental issues. In this respect, it is interesting to look at the evolution of Mercosur and the La Plata basin regime in conjunction with respect to the governance of the Guaraní aquifer. In the same time period when scientific knowledge on the transboundary dimensions of the aquifer consolidated, civil society interest in how this shared and valuable underground water resource should be managed also grew. Initially, the links that civil society organisations had to the Mercosur Parliament seemed promising as an avenue to put their concerns on regional agendas. Yet, eventually the Southern Cone governments decided that the governance of the aquifer was not to be addressed in Mercosur and instead gave preference to the technical and less accessible forums of the La Plata basin regime. Overall, with respect to the La Plata basin civil society organisations have had some influence in stopping, downscaling or delaying large-scale infrastructure and hydropower projects in the basin through the use of "outside strategies" (Uhlin 2011, 852), that is well-organised protests against particular issues. Yet, there are no institutionalised channels which would allow for civil society input in shaping regional environmental cooperation. Instead, governments have focussed on a more technical and scientific perspective that leaves out the often controversial social and political questions surrounding the governance of water in the basin. A focus on technical and scientific aspects is also evident in the case study on migratory species. Here conservation networks often with links to large international NGOs have gained considerable influence through the relatively open institutional structures of the CMS. By working together with governments in the framework of the CMS they have succeeded in promoting regional environmental cooperation and shaping its implementation.

None of the cases demonstrate institutionalised channels for access to decision-making open to all interested civil society organisations, and on the whole, the analysis indicates that governments grant access to international and regional institutions on a selective basis. As a result, the position of national governments becomes central in determining which issues, framings and proposed solutions make it onto political agendas and which civil society organisations have access to regional and international institutions. This is important because it shapes how the objectives of regional environmental cooperation and the proposed solutions to shared environmental concerns are defined. To illustrate this point, it is helpful to consider the two contrasting models of environmental governance in Latin America highlighted by de Castro et al. (2016, 9–11). The first model relies on market-based mechanisms and technological innovations consistent with the "green economy" paradigm that is globally dominant in order to achieve efficient and sustainable use of natural resources and pathways to development. This is prevalent amongst policy circles of most Latin American governments. Opposing this is a radically different model captured under the label of "buen vivir" (good living). Drawing on indigenous ideas this includes alternative conceptions of nature and relations between humans and nature. This view is highly critical of the hegemonic capitalist model which it regards as the origin of environmental problems and injustice. Consequently, proponents of this view argue that unequal power relations need to be addressed in order to solve socio-environmental problems. Moreover, this view is highly critical of the development model adopted by South American governments and the reliance on resource exploitation.[1] In practice, it is not always evident to divide environmental groups according to the two models and governments have also adopted elements of both. It is therefore helpful to regard the two models as opposing ends of a spectrum where one end is characterised by the acceptance of market-based mechanisms consistent with green economy approaches and a willingness to work within existing institutions and dominant policy approaches while the other end advocates radical changes to the capitalist system and rejects the notion of development and seeks instead to move towards "post-extractivist" societies. There are considerable variations in between those two ends, but nevertheless the two models are helpful for understanding the very different approaches and proposed solutions to environmental concerns in South America.

In the case of migratory species, civil society organisations have gained a relatively high level of access, but it has to be noted that the organisations involved are a very particular type of organisation. They are internationally well-connected and relatively well-resourced conservation networks often linked to large international NGOs. They work within the existing institutions and tend to frame their work more as technical advice and not as questions of justice or rights. On the whole, they are closer to the green economy model of the spectrum outlined above. These networks have been able to promote regional environmental cooperation in the framework of the CMS "from the inside", but their influence is also limited to the question of species protection and does not extend to regional integration or other environmental concerns more generally. In addition, due to their international networks, they have also been able to generate their own funding and this makes them attractive partners for governments. As noted by some studies from the 1990s when neoliberal restructuring had drastically reduced the role of the state, some NGOs in Latin America and Asia have taken on quasi-governmental functions benefitting from global networks and funding from Northern partner organisations which is independent of governments (Fairman and Ross 1996, 43–44; Jacobson and Brown Weiss 1998, 533–534). As Raustiala (1997) points out in relation to international environmental institutions, the relationship between NGOs and governments is not a zero-sum game where governments lose power if NGOs gain in influence. On the contrary, governments can benefit from NGO participation if NGOs have expertise and resources that they can make available to governments in order to support them in decision-making or implementation. At the same time governments monitor carefully which organisations gain how much access. In some international environmental negotiations for example NGOs have to be deemed "qualified" to address a particular subject and this can act as a criterion to exclude groups. This applies to the CMS. In this case study governments can thus rely on the resources and expertise provided by NGOs and this allows them to demonstrate their commitment to this particular global environmental convention, but they still retain control over the groups that have access to decision-making processes and implementation.

A comparison of the cases examined in this book suggests that four factors stand out as being particularly important in terms of determining whether and to what extent civil society organisations can shape the nature of regional environmental cooperation. First, some institutional frameworks are more open to civil society participation than others and the CMS and Mercosur offered comparatively more access even if this also remains limited and very much mediated by governments. Second, the focus and framing of civil society matters and approaches that are framed in technical terms and in line with the green economy approach are more likely to gain access to regional or international institutions. Third, groups that are able to offer resources and expertise to governments are more likely to be able to promote and shape regional environmental cooperation "from the inside". Finally, it seems that the saliency of an issue also matters and governments are more likely to grant access for issues with a low level of political contestation as is the case for species protection.

None of these findings are unusual or typical only for South America. Other studies have noted that NGOs have more influence in international environmental negotiations if their arguments do not contradict dominant discourses and if there are lower levels of contention over economic interests (Betsill 2008, 201–202). Previous studies have also found that the influence of an epistemic community increases if there are ways of accessing decision-makers more easily and if the objectives of the epistemic community correspond to existing norms and are not too disruptive. Moreover, it helps if the issue at stake is seen as scientific and technical (Cross 2013, 145). With respect to regional environmental cooperation Elliott (2012, 52) outlines how ASEAN governments have selectively included NGOs in regional environmental governance but without creating wider participation mechanisms. In relation to policy processes in South American politics previous studies have found that although progressive governments have been more open to pressure from below and introduced some more possibilities for civil society participation during the 2000s, this remains uneven and selective (Cannon and Kirby 2012, 196–197) with governments retaining their central position.

Even if these findings are therefore not unusual, they are important because they shape the nature of regional environmental cooperation in particular ways. In the Southern Cone it is thus striking that regional environmental cooperation does not address many of the region's most important transboundary and shared environmental concerns for example in relation to the expansion of intensive agriculture or forestry plantations and the pulp industry notwithstanding the interest that some civil society groups have demonstrated in those issues. Moreover, the findings highlight the need for further and more systematic research on what potential endogenous drivers for environmental cooperation in different regions of the South are and which drivers succeed in promoting cooperation. The findings from the book suggest that if regional environmental cooperation appears to be much more driven by external actors than domestic ones, this does not necessarily mean that domestic actors with an interest in promoting regional environmental cooperation do not exist. Instead, the lack of involvement of endogenous drivers could also reflect the sidelining of interested civil society organisations. This is an important consideration and avenue for further research in cases where endogenous drivers for regional environmental cooperation appear to be absent.

Finally, the prominent role that exogenous drivers play and the unequal influence of endogenous drivers in environmental cooperation in regions of the South also raises important questions regarding the legitimacy and accountability of cooperation. This was not a focus of this book, but it is an increasingly important consideration in environmental cooperation that speaks to the ongoing research programme on earth system governance (Biermann and Gupta 2011; Biermann et al. 2010). Legitimacy and accountability are both concepts which describe norms and standards on the one hand and relations between different actors on the other. Accountability provides a link between those who are held accountable and those who have the right to hold to account according to a particular standard of behaviour (Biermann and Gupta 2011, 1857). Legitimacy relates to the acceptance and justification of authority and equally relates to accepted rules or principles and standards of behaviour. Again, there is a relational element and a crucial question is who regards something, in this case regional environmental cooperation, as legitimate (Biermann and Gupta 2011, 1858). Accountability and legitimacy are important as values on their own, but also affect the effectiveness of environmental governance as cooperation arrangements can be expected to be more effective if they are seen as legitimate and accountable (Biermann et al. 2010, 286–287).

Theoretically, in a strictly intergovernmental process governments are accountable to their voters and this constitutes the main source of legitimacy. International bureaucracies, such as the CMS Secretariat or the CIC in the La Plata basin regime gain legitimacy through the mandate from the governments who are their principals, but this leads to very long lines of accountability which have been questioned (Biermann et al. 2010, 286). The involvement of actors from outside a particular region further complicates lines of accountability and raises more questions regarding the legitimacy of cooperation. It is therefore not at all clear who funders are or should be accountable to: their home constituencies or the institutions who provide the funding; the recipient governments; or the people whose daily lives are affected by projects with external funding? Similar questions can be raised with respect to NGOs that engage in cooperation in a particular region, but receive a large part of their funding from sources outside the region. Concerns like these can contribute to the public distrust of external funders.

In order to shed some light on these questions it is helpful to draw on the notion of public power outlined by Macdonald (2008) in her study of global stakeholder democracy. Macdonald argues that non-state actors should be regarded as agents of public power if they constrain the autonomy of a population of individuals. This can happen by influencing the process of regulative social norm building or through the imposition of material constraints. While any political actor can have influence in the production and maintenance of autonomy-constraining regulative norms, this becomes problematic and therefore requires democratic legitimation only if this influence is higher compared to other actors. In regional environmental cooperation this would mean that democratic legitimacy is lacking if non-state actors such as foreign donors or NGOs have disproportionate influence in shaping regional norms. However, some norms are not strong enough to impose constraints. Arguably, this also applies to the Southern Cone cases where most of the agreements and declarations that make up formal cooperation are not legally binding and often lack specificity. In relation to the second aspect of public power, the imposition of material constraints, Macdonald outlines that external funders also need to make distributional decisions in allocating the available resources and according to her this can also be a constraint because individuals who would have benefitted had other priority decisions been taken, lose out. The examples provided by Macdonald are very different from the marginal forms of regional environmental cooperation in the Southern Cone, and in particular it can be debated to what extent environmental cooperation in the Southern Cone constrains the autonomy of individuals or how much impact it has on the lives of citizens. Nevertheless, these are important considerations especially when studying regions where external actors are particularly influential in regional environmental cooperation. While it has been argued that the regional level has the potential to become a more legitimate scale for environmental cooperation compared to the global level (Balsiger and VanDeveer 2012, 3; Elliott and Breslin 2011, 8–10), this is thus by no means automatic. The processes leading to environmental cooperation in different regions and the roles of external drivers therefore deserve attention in further studies.

REGIONAL INTEGRATION AND REGIONAL ENVIRONMENTAL COOPERATION: TWO SEPARATE PROCESSES

A second important finding that emerged from studying the processes of regional environmental cooperation in the Southern Cone is that regional integration processes led by governments are not necessarily a driver for regional environmental cooperation, but on the contrary, they can also present significant constraints. This is perhaps more unexpected as it is very different from the European experience, but in other parts of the world too, it has sometimes been observed that environmental cooperation grows out of broader regional integration processes (Elliott and Breslin 2011, 15; Siegel 2016a, 719–720). This observation raises important questions about the nature of regional cooperation in South America as well as the validity of the EU as a reference point for regional (environmental) cooperation elsewhere.

It is notable that the most robust case, cooperation on migratory species, has taken place outside the existing regional regimes and in parallel, but largely separate from regional integration processes in the framework of Mercosur or Unasur and the La Plata basin regime. This is significant given that over the last decade South American governments have repeatedly declared their commitment to further regional cooperation going beyond trade and economic questions in order to address also domestic needs. Yet, as outlined in Chap. 3 the mandate of Mercosur's environmental institutions has been extremely limited, curtailing the ability of transnational networks of government officials to become a driver for regional environmental cooperation and both Mercosur and the La Plata basin regime have posed obstacles to civil society participation, sidelining yet another potential driver. Overall, regional integration and regional environmental cooperation appear to be two separate processes and it almost seems that regional environmental cooperation in the Southern Cone has increased over the last two decades in spite of rather than because of government-led regional integration.

Although on paper Mercosur has a relatively high level of institutionalisation with respect to environmental concerns, the inconsistency in environmental topics addressed and the weak position of Mercosur's environmental institutions suggest that environmental concerns made it onto Mercosur's agenda mostly for other political reasons rather than out of a desire to address a particular shared environmental problem. The Southern Cone governments used Mercosur environmental declarations as a way to demonstrate a commitment to international norms while European funders supported environmental cooperation in Mercosur as part of a package promoting EU-style integration in other regions. Moreover, the processes of regional integration, determined very much by the presidential executives, have worked in such a way that potential drivers for regional environmental cooperation, notably regional civil society networks and transnational government networks have faced significant constraints in their ability to work on shared and transboundary environmental concerns.

As Slaughter (2004) has shown transnational government networks between regulators, judges and legislators have significantly increased in scale and scope. Members of such networks work for national governments, but in addition they also embody their professional interests and norms and may work towards these. Transnational government networks seeking to address transboundary environmental concerns are important potential drivers for regional environmental cooperation. Indeed, studies have shown that in particular regulators, that is appointed top officials and civil servants, exist and work on environmental concerns in various regions of the world. Regional organisations such as the Association of Southeast Asian Nations (ASEAN) (Elliott 2012), Mercosur (Hochstetler 2003, 2005, 2011) or the North American Free Trade Agreement (NAFTA) (Slaughter 2004) have all led to networks on environmental concerns between high-level officials at the ministerial level as well as lower-level civil servants. In the case of ASEAN the formation of such transgovernmental networks was the result of a deliberate strategy of the ASEAN member states (Elliott 2012, 49). Moreover, the South Asian Cooperation for Environmental Protection (Matthew 2012, 113; Swain 2011), the Senior Officials Meeting for Environmental Cooperation in Northeast Asia (Elliott 2011, 67) the Commission for the Forests of Central Africa or the African Ministerial Conference on the Environment (Compagnon et al. 2011) are all examples of transnational networks of government officials in less institutionalised frameworks. Such networks can promote regional environmental cooperation by exchanging ideas, techniques and experiences, and offering training or technical assistance, in particular for weaker member countries. They can also work on the harmonisation of laws and regulations (Slaughter 2004, 51–52).

Yet, the analysis of regional environmental cooperation in the Southern Cone presented in the book demonstrates that there are also significant limitations to the influence of such transnational government networks. In South America regional cooperation is very much driven by the presidential executives who determine the agendas, priorities and outcomes of regional cooperation and who are reluctant to delegate power and autonomy to regional institutions that could challenge their own power (Malamud 2015, 172; Malamud and Dri 2013, 234). In the case of Mercosur, governments have not given the environmental institutions of the regional organisation the mandate to address several of the region's most important and transboundary environmental concerns as examined in Chap. 3. Moreover, sensitive issues have frequently been withdrawn from Mercosur's environmental agenda and environmental cooperation remains heavily dependent on external funding which brings its own challenges as discussed above. The mechanisms of regional cooperation within Mercosur have therefore severely constraint the ability of transnational government networks to drive environmental cooperation. Overall, the relevance of Mercosur for regional environmental cooperation in the Southern Cone has declined over time while the more recently created regional organisations Unasur and Alba do not have any institutions dedicated specifically to the environment. On the other hand, over the past decade the regional infrastructure initiative IIRSA has become a central element of regional cooperation, but the initiative is highly contested not least due to socio-environmental concerns. However, channels for civil society participation in relation to those issues are only weakly developed or ineffective. Finally, as discussed in the previous section, the processes of regional cooperation have also worked in such a way that some civil society organisations with an interest in addressing common environmental problems from a regional perspective have not been able to exert much influence.

On the whole, government-led regional integration processes in South America have therefore not become a driver for regional environmental cooperation. On the contrary, they have presented important constraints to this. This challenges some assumptions about the nature of regional cooperation under progressive governments as well as the validity of the EU as a model for regional cooperation elsewhere. As set out in Chap. 3, with the election of a wave of leftist governments a shift in the nature and logic of regional cooperation has taken place as domestic concerns and in particular social objectives became more important while trade and economic objectives lost their centrality as the main drivers for regional integration. This has been accompanied by a focus on regional identity and unity as well as autonomy from external influences and the US in particular. For many on the left these developments sparked the hope for fundamental changes in both domestic and regional politics. Chodor (2015, 149) for example notes that this "opens spaces for radical politics where before they did not exist". However, the findings from the book clearly demonstrate the limitations for radical transformations at least as far as socio-environmental issues are concerned as it is very evident that government-led regional integration processes under progressive governments have not become frameworks for strengthening radically different or alternative approaches to nature and the environment. On the contrary, the findings suggest that more radical and critical approaches focussing on environmental justice and rights have been less influential in shaping regional environmental cooperation than those working within the framework of market mechanisms and seeking technical solutions. From the point of view of environmental governance, regional cooperation appears much more marked by continuity than by radical change confirming the assessment of Riggirozzi and Tussie (2012, 184) that "post-neoliberal" regionalism is "not a rupture with the past, but an evolution, shaped by legacies of past development trajectories, pragmatism, ad hoc policy-making and responses to global and regional politics". Continuity is also evident in the way decisions are taken at the regional level with presidential executives continuing to dominate decision-making as noted above.

With the election of a centre-right government in Argentina in 2015, the controversial impeachment of Dilma Rousseff in Brazil in 2016 and political and economic crises in Brazil and Venezuela, regional cooperation is set to take yet another turn with a renewed emphasis on free trade and good relations with the US and Europe rather than regional autonomy and social concerns. Moreover, in August 2016 a stagnating Mercosur was weakened further when Venezuela under President Maduro announced it had taken over the presidency of Mercosur as foreseen by the alphabetical order, but this was not accepted by Argentina, Brazil and Paraguay while Uruguay attempted to retain a middle-ground position. This clearly shows the difficulties of continuing regional cooperation when this is strongly determined by the executives of the respective countries and these find themselves on opposing ends of the political spectrum. At the same time, civil society is generally in a much stronger position than in the early 1990s at the start of the period analysed here, and civil society demands to address socio-environmental concerns will not disappear. How socio-environmental concerns are addressed and how contestations play out in domestic and regional politics as the region shifts to the right therefore remains a topic for further research.

Finally, the finding that government-led regional integration processes can be drivers as much as obstacles to regional environmental cooperation, and conversely, that regional environmental cooperation also happens outside of government-led integration processes, is also relevant beyond South America. Throughout the study of regionalism, the EU has generally been regarded as the main reference point and "the standard model for regional integration" (Börzel and Risse 2012, 197). In the European case regional integration significantly strengthened environmental cooperation, but as outlined above in South America regional integration processes have taken a very different path so that the analytical value of the EU model is very limited. On the contrary, the case studies examined in the book demonstrate the importance of looking beyond the very narrow vision of "EU-like" regional cooperation as this would have meant overlooking both, cooperation in the framework of the La Plata basin regime and in the framework of the CMS.[2] This also challenges the often implicit assumption that "successful" regional integration will resemble the EU (Elliott and Breslin 2011, 1) as Mercosur, the regional organisation that was at one time regarded as most similar to the EU globally, in fact turned out to be less and less relevant for environmental cooperation in the Southern Cone over the time period examined in the book. Further research may benefit from carefully scrutinising the nature of regional integration and its relationship with regional environmental cooperation. Moreover, examining the environmental agendas of regional organisations taking into account also the environmental issues that they do *not* address and the reasons for this may also generate important insights. Finally, comparing different non-European cases with each other may be more helpful than comparisons with the EU whose relatively strong forms of regional environmental cooperation were developed in a comparatively favourable economic and political context that is very different from many regions of the South as discussed in the final part of this chapter.

THE MARGINALITY OF REGIONAL ENVIRONMENTAL COOPERATION

A third key finding that emerged from the analysis is that conflicting objectives between the need for social and economic development on the one hand, and environmental protection on the other are important to account for the continuing marginality of regional environmental cooperation. This conflict of political priorities or objectives can be traced to the development strategy adopted by governments and this in turn is closely related to the region's position in the global political economy. Chapter 2 has outlined the evolution of approaches to development since the early 1990s. Although regional environmental cooperation has increased in quality and quantity in that time period it has remained marginal as governments have for good reasons focussed on economic and social development. Increasing natural resource exploitation has been an important element of development strategies under both neoliberal governments and progressive governments. In the context of the commodity boom neo-extractivist development strategies allowed governments to maintain a fragile balance between keeping domestic and international economic elites linked to the export sector content while implementing social programmes and responding to important development needs. However, intensive resource exploitation also has significant socio-environmental impacts and in this approach to development environmental concerns have therefore often ended up in opposition to developmental concerns. In particular, the progressive governments of the Andean countries of Ecuador and Bolivia have very clearly justified increased resource exploitation by presenting their political options as a choice between environmental protection or social development (Siegel 2016b). This is important in order to understand why international environmental norms often seem to be adopted by South American governments only superficially, but not fully internalised. As Risse and Ropp (2013, 21) found in relation to human rights norms, the existence of persuasive counter-narratives can be an important factor reducing social pressure on governments to comply with norms. Poverty is a central concern for many South American citizens and governments and during the 2000s the commodity boom and neo-extractivist development strategies have succeeded in making some important social achievements possible. Therefore, it is not surprising that if there is a perception that there is a trade-off between environmental protection and social concerns, the latter win the upper hand.

However, this conflict of objectives or norms between environment and development is very much linked to the region's position in the global political economy. The export of natural resources from South America to other parts of the world is of course not a new phenomenon or a strategy invented by governments over the last 25 years. South America's natural resource wealth has determined the region's position in the global political economy for over half a millennium (Edwards and Roberts 2015, 19). European powers forced South America into the global economy as a provider of primary commodities during the colonial period, and since then, resources have generated enormous wealth for those who controlled them (Galeano 1973). This had devastating social and environmental impacts at the time, but it also contributed to the economic and political marginalisation of South America which lasted beyond independence. As Latin American dependency theorists have argued in great detail in the 1970s, the periphery countries in the South remained subordinate and dependent on the core in the North (Cardoso and Faletto 1979). The return to export-driven growth in the 1990s which is continued under the neo-extractivist model, has reinforced the position of the Southern Cone and South America as a whole as providers of primary commodities in global markets (Green 1999; Murray 1999). Since the turn of the millennium the rise of China has offered alternative sources of investment and trade relationships for South American countries, making the region less dependent on the global North (Fernández Jilberto and Hogenboom 2010; Hochstetler 2013, 39) and therefore provided the region with more policy autonomy and a greater margin for manoeuvre. However, it has also extended the region's resource dependency which is evident for example in the soybean sector (Edwards and Roberts 2015, 20). In recent years declining demands from China leading to falling commodity prices have already demonstrated the economic vulnerability of this development strategy. The search for more sustainable development models is thus made more difficult by South America's longstanding position in the global political economy.

This relationship between regional environmental cooperation and strategies for development in the context of the global political economy is important for South America and beyond. For those interested in strengthening regional environmental cooperation in South America how this tension can be resolved is a central question. In this respect, it is important to point out that environment versus development is a false dichotomy in the sense that increased resource exploitation and environmental degradation are closely linked to developmental concerns such as health problems, availability of clean water, conflicts over resource exploitation or inequalities in terms of benefits and burdens of resource exploitation which frequently reflect other social, economic and political inequalities. Development strategies that ignore socio-environmental concerns are therefore likely to reinforce old developmental concerns and/or create new ones. The current more modest economic outlook could drive environmental concerns further down political agendas and limit resources available to address them, but it could also present an opportunity in terms of developing more sustainable solutions to development for example if energy efficiency and renewable energy were prioritised (Edwards and Roberts 2015, 185). However, due to the linkages of the global economy, it is also clear that South American countries stand little chance of achieving more sustainable development without the support of those buying and consuming the region's natural resources. Overall, the book therefore highlights the importance of analysing regional environmental cooperation in the wider context of strategies for sustainable development and the global political economy. Given that regional environmental cooperation is also a marginal process in many other regions of the South, it would be worth investigating how these factors shape environmental cooperation elsewhere. Even if domestic politics are no doubt often complex and unfavourable for stronger regional environmental cooperation, this cannot be regarded as something that only concerns governments and citizens in the South, but is as much linked to politics and choices made in the North.

NOTES

1. These different approaches to environmental governance reflect some of the issues already raised in debates on the meaning of "sustainable development" during the 1990s when divisions between those advocating stronger and those arguing for weaker interpretations of the concept became evident (Svampa 2015, 27).

2. Elliott and Breslin make a very similar point when they note that there are frequently discussions about the prospects of regionalism in East Asia although in fact there are already many existing forms of regionalism, but they do not resemble EU style regionalism (Elliott and Breslin 2011, 1).

REFERENCES

Balsiger, Jörg, and Stacy D. VanDeveer. 2012. Navigating Regional Environmental Governance. *Global Environmental Politics* 12 (3): 1–17.

Betsill, Michele M. 2008. Reflections on the Analytical Framework and NGO Diplomacy. In *NGO Diplomacy—The Influence of Nongovernmental*

Organizations in International Environmental Negotiations, ed. Michele M. Betsill and Elisabeth Corell. London: MIT Press.

Biermann, Frank, and Aarti Gupta. 2011. Accountability and Legitimacy in Earth System Governance: A Research Framework. *Ecological Economics* 70 (September): 1856–1864.

Biermann, Frank, Michele M. Betsill, Joyeeta Gupta, Norichika Kanie, Louis Lebel, Diana Liverman, Heike Schroeder, Bernd Siebenhüner, and Ruben Zondervan. 2010. Earth System Governance: A Research Framework. *International Environmental Agreements: Politics, Law and Economics* 10 (4): 277–298.

Börzel, Tanja A., and Thomas Risse. 2012. When Europeanisation Meets Diffusion: Exploring New Territory. *West European Politics* 35 (1): 192–207.

Cannon, Barry, and Peadar Kirby. 2012. Civil Society—State Relations in Left-Led Latin America: Deepening Democratization? In *Civil Society and the State in Left-Led Latin America—Challenges and Limitations to Democratization*, ed. Barry Cannon, and Peadar Kirby. London: Zed Books.

Cardoso, Fernando Henrique, and Enzo Faletto. 1979. *Dependency and Development in Latin America*. Berkeley and Los Angeles: University of California Press.

Chodor, Tom. 2015. *Neoliberal Hegemony and the Pink Tide in Latin America*. Houndmills: Palgrave Macmillan.

Compagnon, Daniel, Fanny Florémont, and Isabelle Lamaud. 2011. Sub-Saharan Africa—Fragmented Environmental Governance without Regional Integration. In *Comparative Environmental Regionalism*, ed. Lorraine Elliott, and Shaun Breslin. Oxon: Routledge.

Connolly, Barbara. 1996. Increments for the Earth: The Politics of Environmental Aid. In *Institutions for Environmental Aid*, ed. Robert O. Keohane and Marc A. Levy. London: MIT Press.

Cross, Mai'a K. Davis. 2013. Rethinking Epistemic Communities Twenty Years Later. *Review of International Studies* 39 (01): 137–160.

De Castro, Fabio, Barbara Hogenboom, and Michiel Baud. 2016. Introduction: Environment and Society in Contemporary Latin America. In *Environmental Governance in Latin America*, ed. Fabio de Castro, Barbara Hogenboom, and Michiel Baud. Basingstoke: Palgrave Macmillan.

Edwards, Guy, and J. Timmons Roberts. 2015. *A Fragmented Continent—Latin America and the Global Politics of Climate Change*. Cambridge, MA: MIT Press.

Elliott, Lorraine. 2011. East Asia and Sub-Regional Diversity—Initiatives, Institutions and Identity. In *Comparative Environmental Regionalism*, ed. Lorraine Elliott, and Shaun Breslin. Oxon: Routledge.

Elliott. 2012. ASEAN and Environmental Governance: Strategies of Regionalism in Southeast Asia. *Global Environmental Politics* 12 (3): 38–57.

Elliott, Lorraine, and Shaun Breslin. 2011. Researching Comparative Regional Environmental Governance—Causes, Cases and Consequences. In *Comparative*

Environmental Regionalism, ed. Lorraine Elliott, and Shaun Breslin. Oxon: Routledge.

Fairman, David, and Michael Ross. 1996. Old Fads, New Lessons: Learning from Economic Development Assistance. In *Institutions for Environmental Aid*, ed. Robert O. Keohane and Marc A. Levy. London: MIT Press.

Fernández Jilberto, Alex E., and Barbara Hogenboom. 2010. Latin America and China—South-South Relations in a New Era. In *Latin America Facing China: South-South Relations Beyond the Washington Consensus*, ed. Alex E. Fernández Jilberto and Barbara Hogenboom. New York: Berghahn Books.

Galeano, Eduardo. 1973. *Open Veins of Latin America*. New York: Monthly Review Press.

Green, Duncan. 1999. A Trip to the Market: The Impact of Neoliberalism in Latin America. In *Developments in Latin American Political Economy—States, Markets and Actors*, ed. Julia Buxton, and Nicola Phillips. Manchester: Manchester University Press.

Hochstetler. 2003. Fading Green? Environmental Politics in the Mercosur Free Trade Agreement. *Latin American Politics and Society* 45 (4): 1–32.

Hochstetler. 2005. Race to the Middle: Environmental Politics in the Mercosur Free Trade Agreement. In *Handbook of Global Environmental Politics*, ed. Peter Dauvergne. Cheltenham: Edward Elgar.

Hochstetler. 2011. Under Construction—Debating the Region in South America. In *Comparative Environmental Regionalism*, ed. Lorraine Elliott and Shaun Breslin. Oxon: Routledge.

Hochstetler. 2013. South-South Trade and the Environment: A Brazilian Case Study. *Global Environmental Politics* 13 (1): 30–48.

Jacobson, Harold K., and Edith Brown Weiss. 1998. Assessing the Record and Designing Strategies to Engage Countries. In *Engaging Countries—Strengthening Compliance with International Environmental Accords*, ed. Edith Brown Weiss and Harold K. Jacobson. London: MIT Press.

Kulauzov, Dora, and Alexios Antypas. 2011. The Middle East and North Africa—Sub-Regional Environmental Cooperation as a Security Issue. In *Comparative Environmental Regionalism*, ed. Lorraine Elliott, and Shaun Breslin. Oxon: Routledge.

Lewis, Tammy L. 2016. *Ecuador's Environmental Revolutions: Ecoimperialists, Ecodependents, and Ecoresisters*. Cambridge, MA: MIT Press.

Macdonald, Terry. 2008. *Global Stakeholder Democracy: Power and Representation Beyond Liberal States*. Oxford: Oxford University Press.

Malamud. 2015. Interdependence, Leadership and Institutionalization: The Triple Deficit and Fading Prospects of Mercosur. In *Limits to Regional Integration*, ed. Søren Dosenrode. Farnham: Ashgate.

Malamud, Andrés, and Clarissa Dri. 2013. Spillover Effects and Supranational Parliaments: The Case of Mercosur. *Journal of Iberian and Latin American Research* 19 (2): 224–238.

Matthew, Richard. 2012. Environmental Change, Human Security, and Regional Governance: The Case of the Hindu Kush/Himalaya Region. *Global Environmental Politics* 12 (3): 100–118.

Murray, Warwick E. 1999. Natural Resources, the Global Economy and Sustainability. In *Latin America Transformed—Globalization and Modernity*, ed. Robert N. Gwynne and Cristobal Kay. London: Arnold.

Raustiala, Kal. 1997. States, NGOs, and International Environmental Institutions. *International Studies Quarterly* 41 (4): 719–740.

Riggirozzi, Pía, and Diana Tussie. 2012. Postlude. In *The Rise of Post-Hegemonic Regionalism—The Case of Latin America*, ed. Pía Riggirozzi, and Diana Tussie. Dordrecht: Springer.

Risse, Thomas, and Stephen C. Ropp. 2013. Introduction and Overview. In *The Persistent Power of Human Rights—From Commitment to Compliance*, ed. Thomas Risse, Stephen C. Ropp, and Kathryn Sikkink. Cambridge: Cambridge University Press.

Siegel, Karen M. 2016a. Can Regional Cooperation Promote Sustainable Development? In *The Palgrave Handbook of International Development* ed. Jean Grugel and Daniel Hammett. Basingstoke. Palgrave Macmillan.

Siegel, Karen M. 2016b. Fulfilling Promises of More Substantive Democracy? Post-Neoliberalism and Natural Resource Governance in South America. *Development and Change* 47 (3): 495–516.

Slaughter, Anne-Marie. 2004. *A New World Order*. Princeton: Princeton University Press.

Svampa, Maristella. 2015. ¿El Desarrollo En Cuestión? Algunas Coordenadas Del Debate Latinoamericano. In *El Desarollo En Disputa - Actores, Conflictos Y Modelos de Desarollo En La Argentina Contemporánea*, ed. Maristella Svampa. Los Polvorines: Ediciones Universidad Nacional de General Sarmiento.

Swain, Ashok. 2011. South Asia, Its Environment and Regional Institutions. In *Comparative Environmental Regionalism*, ed. Lorraine Elliott, and Shaun Breslin. Oxon: Routledge.

Uhlin, Anders. 2011. National Democratization Theory and Global Governance: Civil Society and the Liberalization of the Asian Development Bank. *Democratization* 18 (3): 847–871.

Index

A
Accountability, 161
Accountability politics, 19, 135
Agribusiness, 46, 50, 51, 65, 78, 79, 81, 83
Agriculture
  production, 45, 49, 80, 109, 130, 141
Alianza Bolivariana para los Pueblos de nuestra América (Alba)-Bolivarian Alliance for the Peoples of our America, 14, 64, 77, 78, 83, 165
Amazon, 3, 4, 37, 40, 78, 92
Andean countries, 47, 168
Andean flamingos, 130, 134, 136, 142
Argentina, 3, 6, 11–13, 15, 36, 37, 41–43, 46, 47, 51, 64, 65, 67, 69, 70, 80, 91, 92, 94, 96–98, 100–106, 115, 127, 131, 133, 136, 140–142, 145, 166
Association of Southeast Asian Nations (ASEAN), 160, 164
Aves Argentinas, 130, 140, 142

B
Birdlife International, 130, 136, 141, 142

Bolivia, 12, 43, 46, 48, 50, 80, 91, 92, 100, 101, 103, 115, 126, 131, 136, 145, 168
Bolsa Familia, 47, 81
Bonn Convention. *See* Convention on Migratory Species
Boomerang strategy, 99
Botnia, 98
Brasiguayos, 82
Brazil, 82, 83, 91, 92, 94, 96, 98–103, 126, 128, 140, 141, 145, 166
Buen vivir, 158

C
Capitalism, 1, 4, 9, 11, 14, 23, 34, 35, 40, 42, 44, 45, 48, 49, 54, 63, 66, 76–78, 80, 92, 94, 98, 105, 108, 109, 114, 123, 141, 145, 153, 163, 165, 167–169
Capitalist economy, 52
Centro de Formación para la Integración Regional (CEFIR)-Centre of Education for Regional Integration, 71
Centro Latinoamericano de Ecología Social (CLAES)-Latin American Centre of Social Ecology, 66–69, 71
Chile, 46, 47, 67, 126, 131–133, 145

K.M. Siegel, *Regional Environmental Cooperation in South America*, International Political Economy Series, DOI 10.1057/978-1-137-55874-9

Printed by Printforce, the Netherlands